SOLAR ENERGY UTILIZATION

SOLAR ENERGY UTILIZATION

Tim Michels

President
Londe • Parker • Michels, Inc.
Energy Consultants
St. Louis, Missouri

VNR **VAN NOSTRAND REINHOLD COMPANY**
NEW YORK   CINCINNATI   ATLANTA   DALLAS   SAN FRANCISCO
LONDON   TORONTO   MELBOURNE

Van Nostrand Reinhold Company Regional Offices:
New York   Cincinnati   Atlanta   Dallas   San Francisco

Van Nostrand Reinhold Company International Offices:
London   Toronto   Melbourne

Library of Congress Catalog Card Number: 79-15840
ISBN: 0-442-25368-0

Manufactured in the United States of America

Published by Van Nostrand Reinhold Company
135 West 50th Street, New York, N.Y. 10020

Published simultaneously in Canada by Van Nostrand Reinhold Ltd.

15 14 13 12 11 10 9 8 7 6 5 4 3 2 1

Library of Congress Cataloging in Publication Data

Michels, Tim.
    Solar energy utilization.

    Bibliography: p.
    Includes index.
    1. Solar energy. I. Title.
TJ810.M52   333.7   79-15840
ISBN 0-422-25368-0

To the children of the world . . . .
that we not leave our problems
for them to solve.

PREFACE

Energy Conservation is much more important than **Solar Energy Utilization**—especially in the near future. Not needing the energy—ever again—is a much healthier posture than producing the same amount of energy by solar means.

Energy is the foundation upon which ever more complex societies are laid and sustained. Energy is the glue that keeps things apart. As energy becomes more costly and scarce, "things" will have to coalesce—draw nearer. If the rate of energy withdrawal is too fast, there is the risk that the coalescing could become a collapse. This transition **must** be managed in the very near future. The shamefully high level of energy waste in the United States has to end. The options are: to do it smoothly—in an orderly, managed way; or, catastrophically—when supplies suddenly are not there.

Designers have a special responsibility, because they place energy loads on society with the very design of buildings. This throws the real challenge for energy conservation right in the lap of the designers, they have to develop the art of designing the building to provide the comfort conditions rather than relying on mechanical equipment (both the architects and the engineers).

Some architects and engineers have not yet made the connection between design and the energy maintenance costs of a design. Some clients paying the bills for heating and cooling have not made the connection yet either, but they will probably come to the realization sooner than the designers do. It must be understood that the solution is not in developing more efficient mechanical equipment, but in producing more efficient designs. The design professionals need to be re-educated. They are just as trapped by the "promise" of technology as society on the whole is. We cannot "wait until Argonne Labs develops the solar collector we need and we'll just put it on the buildings," as one architect suggested.

There is an attitude that must be developed in order to cope effectively with the twin problems of energy conservation and the proper use of solar energy. That attitude is: To do as much as possible with as little as possible. Most of the problem **can** be handled by design. Another goodly portion can be achieved in the re-education of the client. This is definitely within the province of the designers. This will lie, primarily, in helping the client realize that **he** can no longer afford the financial cost of energy waste. Or, at another scale, the client will have to realize that cheap, speculative buildings might be injurious to society itself. Society, as the client of the designer-advocate, will have to realize that: 1) mass transit is something we cannot afford not to undertake; 2) building more energy producing facilities is not the answer when we are currently wasting 50% of the energy produced; and 3) subsidizing economic profits for industry by absorbing the cost of pollution is ruinous both to our physical health and economic well being —in terms of encouraging the waste of capital resources, such as minerals, water and fuels.

Energy is already one of the more critical variables that must be juggled in the design process. In the future, solutions to specific and general building projects will also have to be solutions within the overall framework of energy consumption and the problems it poses. Designers must be ready to operate within this arena, or they will find that the control they have over the built environment will be further eroded.

A lot of the problems mentioned (and more) are problems which have solutions only in the political arena. Still, designers, as all other citizens, must become informed so that they may fully participate in the debates that are sure to come. **Now is the time.**

ACKNOWLEDGMENTS

This document represents the collation and digestion of some of the vast amount of literature that has been and is being generated on this now popular topic. Only small portions of the contents represent any "original" synthesis of concepts. The purpose of this document is to further the information base for those who wish to pursue the useful application of solar energy.

I have tried to be as accurate as possible in presenting the information I have gathered, but I have not hesitated in offering my own conjecture where information was scanty, unavailable or nonexistent. Omissions and mistakes are of my own doing and I would gratefully request correction from those to whom these errors are apparent.

Credit for much of the will to complete this project and most of the support in moving it through various rewrites into final copy goes to my wife, Wanda.

Special thanks have to be given to several friends who were instrumental in the refinement of ideas and of presentation:

George Brown	Rudd Falconer
Ed Mass	Iain Fraser
Michael Steiner	Ken Labs
Bernard Yudow	Tyrone Pike

Their support at various critical stages ensured the completion of this work. Within the Londe-Parker-Michels, Inc. support network, special thanks go to Jim Barnes, Alice Enders, Vaughn Bradshaw and Diana Dyer for unflagging effort through the trials of final production. I would also like to thank Advertising and Design Services of St. Louis for their professional typesetting efforts.

Over the four and one-half years this book was in development, many others, too numerous to list, also bolstered a flagging author with their kind encouragement. To these I owe heartfelt thanks.

INTRODUCTION

This book is presented in five sections:

1. Energy Conservation
2. The Sun with the Earth
3. Collectors
4. Storage
5. Sizing

Chapters 2-4 present the necessary background solar information so that one can understand what has been done and, generally, why. This information is required to establish an informed judgement. The designer that is more than passingly familiar with these chapters should be very comfortable in specifying systems or in making quick judgements about systems that may be under consideration for a particular project.

Chapter 5 stands on its own as a quick-study chapter for those with little time or a passing, general interest. It is designed to ensure that basic, preliminary design decisions are made that facilitate design development.

"Energy Conservation" has been placed first because it is more important than solar energy utilization.

"The Sun with the Earth" presents the relationship between the sun and the earth in discussing the variations in solar availability.

"Collectors" develops the information needed for a designer to know what kind of solar approach is most appropriate for his needs. It develops the information on both passive and active systems and it helps the designer make a knowledgeable selection among the different systems and components available.

"Storage" discusses the different methods for energy storage, their advantages and disadvantages and their workability with different collector systems. This discussion is aided by examples of existing solar installations. Care has been taken to develop the concepts from generic examples.

"Sizing" discusses the actual parameters of a system that are most important for a designer to understand in approaching the form of a building on which solar energy is to be utilized. This chapter contains the necessary information and examples for a designer to make quick decisions about **how much** of a system he will need and **how** it relates to the building under consideration. This chapter does not require a thorough knowledge or even an acquaintance with Chapters 1-4.

The discussion and examples focus on the residential use of solar energy for many reasons:

1. Residential space heat is most amenable to solar utilization and it accounts for almost $1/5$ of the energy consumed in this country. It is within the control of individuals who will act in their own, enlightened self-interest without waiting for there to be any "energy policy" at any governmental level.

2. There are very few examples or experiments with any other application at markedly different scales.

3. Most people's interest in solar will be in applying it where it can do them the most direct good—their own home. From this point on they may be able to develop a sense for the probable solar solutions at larger scales.

4. Most importantly, the author needed to discuss it within a framework that would have relevance for students and other interested parties.

For these reasons this book has tried to develop those tools directly necessary for solar energy utilization at relatively small scales. The paramount thing that one should keep in mind is that anything built will be a solar collector—whether it is intended to be or not. For anything being built, first make it a **good** passive solar collector (not forgetting the summer), then decide whether other solar devices are required.

April, 1979
Tim Michels

TABLE OF CONTENTS

SOLAR ENERGY UTILIZATION

CHAPTER 1
ENERGY CONSERVATION

GENERAL COMMENTS

Solar Energy, along with other forms of energy such as wind power, ocean thermal gradients and bio-methane generation, is often referred to as an **alternative** source of energy; but all too often these alternative sources of energy are viewed merely as **additional** sources of energy. The distinction being made here is critical to the perception of energy and its use in this technological society.

Currently, it seems that the "energy crisis", as it has been called, is not perceived as an ongoing pro-position: that it is something more than a passing in-convenience. For many, as the current demand for energy grows, solar energy will be developed as an energy supplement.

When solar power is viewed as an **additional** source of energy, there is an underlying assumption that traditional energy sources will continue to pro-vide energy at the current levels. This perspective neglects the fact that fossil fuels resources are finite: no matter what levels of energy they supply they will eventually be exhausted. The question is not one of "if", but "when".[1] Non-renewable resources should be used as fuels only as a bridge to carry us until renewable sources of energy have completely sup-planted them.

As an **alternative** source of energy, solar energy would be increasingly substituted for conventional sources of energy. This attitude toward the use of solar energy would be representative of a people who have realized that they must curb their demand for energy and their past habits of wasting it. As a society places a greater reliance on solar power, the fact that it is a variable source should create an attitude of careful use of energy, since one never knows what the weather brings or how long the supply of stored energy will have to last.

Yet, no matter what is done in implementation of alternative sources of energy, they will have little effect until the waste of energy is curbed. Recent projections suggest that more than 30% of energy presently produced can be conserved through better energy management; and that large portions of this amount, which are lost in the production of energy itself, can be saved by utilizing more efficient ways of producing energy.[2]

It is conservation of energy that will ease the impact of the energy crisis; and solar energy will have its greatest effect when it works with both energy conservation and effective energy utilization. The first steps in reaching this goal is to achieve a turn about in public attitude toward energy consumption and conservation. Everybody consumes energy and everybody will have to cooperate in its conservation. Consumers will have to studiously avoid those products that are part of the "throw away" economy and to adamantly insist on durable goods. The whole routine of conspicuous consumption has to be eliminated. Consumers will have to question why they want something, and whether or not they really need it. This basic change of lifestyle and attitudes must ultimately express itself in the economics of the marketplace. Products for which an artificial demand has been created and goods designed for "planned obsolescence" will be eliminated if there is no profit in them.

These things will eventually have to be done—it's inevitable. There are two mechanisms by which it can occur. The first is that people can come to understand that energy is a precious commodity with environmental consequences and spontaneously try to conserve it. Or, they can continue to consume enormous amounts of energy, and scarcity will drive up the cost so much that energy intensive goods will ultimately price themselves out of the marketplace.

This second alternative will also have severe consequences socially and politically. If it comes about this way the scale of the impact will be magnitudes greater than the "energy crisis" brought about by the 1973 oil boycott. This would cause chaos at all levels of the public sector, resulting in a very disorderly attempt to cope with a problem whose solution **should** have been worked out long ago.

In the scheme of energy conservation, the architect, as an agent for society has both greater opportunities and greater responsibility to effect changes in energy consumption than the average citizen. The impact of the architect should not be underestimated. He can do much just with building design itself to conserve energy and to take advantage of solar energy...passively. In specifying energy delivery systems, he can require that solar energy be utilized to provide low grade (temperatures below the boiling point of water) heat to all thermal processes. Solar energy systems **are** on the market that can readily supply this heat. His design can be such that he anticipates the further development of solar applicability and provides for easy expansion to a solar capability.

Currently many industrial processes require temperatures much higher than can be supplied by inexpensive collector systems, and high grade fuels are used to produce all the heat in these processes. However, solar power can be used to preheat water to its boiling point, and the higher grade fuels will be needed only to boost the temperature to its required level. In the future, more economical ways may be developed for utilizing solar energy as a high temperature energy source; and, at the same time, many processes can probably be redesigned to operate at lower temperatures. The architect will have to keep abreast of such developments to ensure more effective energy utilization: matching the capabilities of the energy source to the temperature requirements of an energy process.

Architects, as professionals, should try to shape policy in the directions that have the greatest potential for reducing energy consumption. Pushing for mass transit would be one good example, but curtailing speculative building is an example much closer to home. Almost by definition, speculative building does not fill an existing demand. It tries to build in anticipation of a real demand in the future, or it builds with the intention of creating a demand by "image." In the first case, it ultimately fills a real need. In the second, however, it is very wasteful; it creates a false need for something like "prestige" office space. This begins a whole cycle of unnecessary erosion of existing building inventory as older buildings are vacated to fill the new developments. This process often leads to the demolition of healthy buildings solely on economic grounds. These buildings that are lost represent a lot of energy in terms of the processing of their materials, their erection and their demolition, not to mention the fact that the new building, going up on the same spot, is probably much more energy intensive. This cycle is brought on artificially by economic pressures for higher returns which demand higher development—with no regard to energy or the environmental costs. If this process were analyzed and a levy were placed on new, speculative construction that represented the cost to the environment, then the cycle would be dramatically slowed. The speculators should not be able to profit from the waste they generate that society has to absorb at great expense.

The structure of the current economic cycle of a building has to be altered. At the present time economic interest in a building by an investing party is limited to seven to ten years. After this period of time, an investor has reaped his maximum return on the building, and it is sold. This promotes an interest in keeping initial cost as low as possible for still higher returns. This often results in equipment and construction of inferior quality because the budget is inadequate, all of which triggers accelerated deterioration of the building. But, someone who is getting rid of a building in a few years is not interested in long-term maintenance costs and simply does not care if the building falls apart after he takes his leave. Since cheaper construction is often synonymous with short-lived buildings—they die early and have to be demolished. This redoubles the energy costs of building and fuels the cycle of demolition and rebuilding.

The profitability in buildings has to be restructured to encourage building for the long term. The rewards should be reaped by those who keep the energy costs lowest over the longest period of time. There seems to be a good correlation between energy consumption and costs of durable goods. Less energy is generally required—in the long run—by equipment that has a long life and only requires low maintenance. Value engineering techniques also suggest that equipment with longer serviceability, although it is more costly at the outset, generally provides savings over the long term in lower maintenance and replacement costs. It must be noted, however, that energy conservation is not synonymous with higher cost. Many energy conservation measures can reduce the total building cost **without** increasing the initial cost. A case study will illustrate this point quite well.

"Glass can account for something like 20 to 30 percent of the energy used for both heating and cooling. This unfavorable impact can be countered by the designer in a variety of ways. One way is to reduce the window area. He can also use shading devices, or better types of glass, reflecting types, or insulating glass.

Everybody also knows that this special glass is a lot more expensive than ordinary glass. Let me give you one example where there is enough data to show that it is cost-effective to use the more expensive glass in the first place.

I'm speaking now about the Toledo-Edison building. The architects along with Libby-Owens-Ford did a detailed computer study of the effects of different glass options on the building's construction and operating costs. They ended up selecting a chromium-coated dual wall insulating glass that involved an additional expense of something like $122,000 for this particular building. That's over ordinary glass, one quarter-inch thick, that they might have used.

The payoff for this selection came in two forms. Initially, they had an offsetting savings of more than the expense of the glass by $1000, namely $123,000 in the heating and cooling equipment, duct work and other equipment that they didn't need to install in the first place. They broke even right there.

Now, the projected energy savings turned out for this particular building to be more than 5000 kilowatt hours a day, assuming an eight-hour working day, a yearly operating savings of—at current dollars—about $40,000."[3]

According to the National Bureau of Standards, one-third of all energy consumption goes into the operation of buildings and 30-50 percent of that energy is wasted.[4] There is definitely room for improvement —especially in new construction. But, if energy conservation depends on new construction, it will be a long time in coming, since buildings are only being replaced at the rate of 2% a year (and maybe it should be less). However, energy savings of 20% in older buildings is possible right now without any remodeling.[5] Simple policy changes like going to daytime janitorial service, and establishing new standards for heating, air conditioning and lighting in non-critical areas can do it. As energy consultants, Londe-Parker-Michels, Inc. has documented cost effective energy conserving measures in existing buildings of 60% or more. Society should strive to implement policies that will promote long-term energy conservation in the building sector.

Architects, as members of the building team, should work with local materials as much as possible. If possible, they should use less energy intensive materials: materials which require less energy in terms of fabrication, transportation, erection and maintenance.

RESIDENTIAL

The most effective way to conserve energy is to reduce the need for energy. Keeping the amount of space that has to be built to a minimum reduces consumption by never needing it. Building less space requires less energy in terms of both materials and maintenance...all other things being equal. The architect will have to work closely with the client to ensure that excess space is curtailed.

Some changes in lifestyle or expectations may be in order. Questions about having a separate room for every function, or a showcase "living room" need to be asked. Conservation, however, **does not** mean people have to lower their standard of living.

The form, or volume, of the building has an important impact on the amount of energy used. If the site is located where winters are severe, then the surface area has to be kept to a minimum, which implies that the plan of the house should be as square as possible, and that the volume should approach that of a cube (Figure 1.1). If the site is in an area where summer comfort is the most important consideration, then a less compact, more linear plan is called for to take advantage of flow-through, or cross, ventilation. This would mean that the floor plan should be open; walls that would disrupt the flow of air through the house should be avoided (Figure 1.2). This points up the fact that, in most climates, outdoor air can pro-

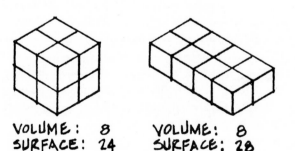

VOLUME: 8　VOLUME: 8
SURFACE: 24　SURFACE: 28

Figure 1.1

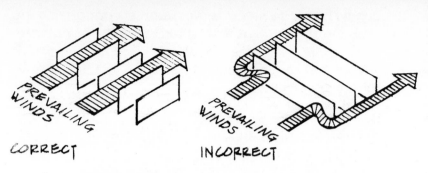

Figure 1.2

vide comfort for substantial periods of the year. Mechanical heating and cooling should be used sparingly. But, during periods when heating and cooling is not required, designs have made it necessary to run fans to supply ventilation. However, this energy for fans can be avoided if the house is designed for natural ventilation from the start.

Once the area required and the basic form have been determined, then the building has to be placed on the site. The site has to be well studied to determine prevailing seasonal winds so that the dwelling can be placed to take advantage of summer breezes and protect itself from winter winds. The topography of the site could be important if it is something besides flat. Where possible, the building should be placed on south facing slopes, rather than in valleys or on north slopes, since valleys and northern slopes generally have colder microclimates. In fact, as far as topography is concerned, the building can be inserted into the slope or even buried underground. Using the earth itself as a protective cover could result in large energy savings in the elimination of all kinds of "finish" materials and in both the heat loss due to transmission through walls and air infiltration from cracks around openings.[6]

The trees on a site are another important consideration in placing the house. Most trees can provide shade for coolness in the summer, but only deciduous trees will allow the house to take advan-

tage of the warmth of the winter sun (Figure 1.3). An analysis of the shading effect of trees is particularly important if use of solar collection is anticipated. Evergreen trees and shrubs are fairly effective as windbreaks in the winter; but they have to be used carefully, or they will interfere with the cooling summer breezes. Summer shading of windows and walls can also be achieved by roof overhangs, porches, screens, fins, etc.; but these shading devices should be designed to allow winter sun penetration to the building surface (Figure 1.4).

All of the above are items that have to be worked together, on the site, to produce a dwelling that takes advantage of the natural conditions. The effect of wind, sun and site on solar control and ventilation in buildings is thoroughly discussed in Victor Olgay's **Design With Climate**. The reader is referred to this source for a more detailed discussion of these points.

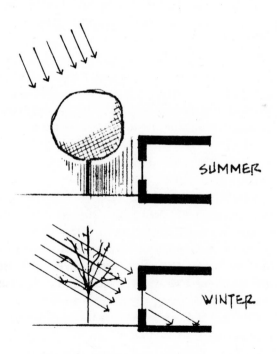

Figure 1.3

7

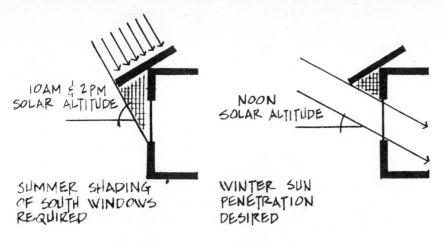

10AM & 2PM SOLAR ALTITUDE

NOON SOLAR ALTITUDE

SUMMER SHADING OF SOUTH WINDOWS REQUIRED

WINTER SUN PENETRATION DESIRED

Figure 1.4

WALLS

In the residential sector, there are two basic construction types: wood frame and masonry. Wood frame construction can be very effective at retarding heat transmission, but it has very little in the way of thermal mass: it does not absorb or hold heat very well. On the other hand masonry construction has relatively high thermal mass, but it is very poor at resisting heat transmission: it conducts heat very well from one side of the wall to the other. In order to utilize the best features of both, a mixture of these construction types is required. If frame construction is used as exterior walls, its best feature—resisting heat loss—is used to good advantage. If masonry walls are used as interior partitions, their thermal mass can be effectively used without their rapid heat loss problem. Such partitions will level out the heating and cooling load of the house. In fact this leveling can probably be used to reduce the size of the heating and air conditioning units because it will help shave peak loads down.[7] The most notable comfort effect would occur where these partitions are placed in south facing rooms with windows. On a sunny winter day, there is often an excessive build up of heat in a south facing room, the masonry wall would work to keep temperatures comfortable by soaking up the heat that would normally have to be vented; later the heat in the walls will be reradiated, as needed, when temperatures drop. In the summer, off peak electrical power can be used during the night to cool the house. The masonry partitions will become "chilled" and effectively cool the house during the day by absorbing the heat. This technique of using interior masonry partitions to store coolness and warmth is part of Passive Solar Design Techniques and is thoroughly discussed in Chapter Three.

WINDOWS

As just pointed out, windows in south walls can turn a room into a solar collector. This brings up the question of the proper use of fenestration. In general a lot more glass is used in houses than is needed for adequate light and view. Though there might be valid aesthetic reasons for large amounts of glass, the glass should be at least used properly. Designing small, operable windows instead of large expanses of fixed plate glass reduces heat loss in winter, heat gains in summer and allows the use of natural ventilation instead of air conditioning. (The "picture window" generally goes with the "showcase" living room.) For winter optimum, there should be less glass on the north walls and more glass on south walls. For summer sun control, the glass on the east and west walls should be minimized, especially on the west; and the southern windows should be shaded to reduce heat gain (Figure 1.4).

In winter, the glass on south walls can provide a substantial amount of heat on sunny days, boosting the performance of energy systems for the house—during the day. However, studies have shown that these large glass areas can be an energy liability in terms of the heat they lose for the 24 hours of the day.[8] There are two ways to respond to this problem of heat loss through windows. One is to reduce the amount of window area so that heat loss becomes insignificant. There are limits, however, codes have set the minimum glass area to around one-tenth of the floor area, half of which has to be ventable;[9] so heat loss from glass will generally be fairly substantial. In addition, if the area of glass is kept to a minimum, the effect of using windows as solar collectors is minimized. The second response is to keep the windows relatively large, to enhance the collector effect, and to devise some insulating shutter arrangement and close the shutters at night (Figure 1.5). Since glass is useless for looking out of at night, there is no reason to leave it exposed, and the insulated shutter would be very effective at reducing heat loss through the glass.

The kind of glazing used can also effect the heat loss. Figure 1.6 is a table of heat transmission coefficients for various arrangements of glass in exterior walls.[10] The comparison between single glass and double insulating glass shows that there is a major improvement in using double insulating glass. By comparison, choosing triple glass instead of double glass produces only marginal improvement. The return in energy savings has to be weighed against the economic costs: could the extra expense of triple glass be better spent somewhere else to achieve more energy savings? By the same token storm windows over single glass produce results equal to the best double insulating glass. So, storm windows would probably be a better economic investment than putting in new insulating glass if one were contemplating remodeling to conserve energy. The important thing to be considered here is: where does

Glass Type	U-Factor (Btu/sq.ft.-Hr-°F)
Single	1.13
Insulating: Double	
3/16" air space	0.69
1/4" air space	0.65
1/2" air space	0.58
Insulating: Triple	
1/4" air space	0.47
1/2" air space	0.36
Storm Windows	
1 to 4" air space	0.56

Figure 1.6

the point of diminishing returns occur? It might well be that a lot more heat loss could be prevented if the same effort, or money, were put into proper weather-stripping than put into insulating glass. Care should be taken in all efforts at energy conservation to ensure that the most good is accomplished with the available resources (often money). In this way the greatest amount of energy will ultimately be saved.

One should be wary of any single item ever being identified as **the** culprit in heat loss. Everything has to work in unison and it has to be seen in proper perspective in order to keep efforts in any one area commensurate with its relative importance.

INFILTRATION

It has been variously estimated that 25-50 percent of the heating and cooling requirements for buildings is a result of infiltration of outside air into the building.[11] The rate of infiltration can be measured in total air changes per hour. Figure 1.7 is a table of air change factors based on openings in the wall.[12] The factors represent how many times the entire volume of air in the house is changed. The fact that this

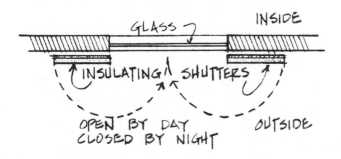

Figure 1.5

Kind of Room or Building	Air Changes per Hour
No windows or exterior doors	0.5
Windows or doors on one side	1
Windows or doors on two sides	1.5
Windows or doors on three sides	2
Entrance Halls	2

Figure 1.7

volume of air is constantly heated up from outside air temperatures, only to be sucked through the wall to the outside again, is responsible for the large heat losses due to infiltration.

These factors are based on whether or not a wall has exterior openings in it or not; and it a is fairly accurate method of estimating infiltration. However, a more exact, detailed technique is based on the total perimeter around the openings in the walls. For this reason the area of glass in a wall can have a substantial effect on the rate of air change; and it is one of the reasons that smaller window areas are suggested for energy conserving buildings. The factors also indicate the importance of weatherstripping or storm windows, since with either of these two items the factors are reduced one-third.

Another way to reduce infiltration is to design buffered entries. Placing such enclosed but unheated areas as garages, mud rooms, entry vestibules or sun porches between doors to the heated part of the dwelling and the outside prevents, or dramatically reduces, the air changes involved when opening and shutting doors.

As mentioned before, placing a dwelling underground practically eliminates unwanted infiltration. But it does introduce a need for positive ventilation. Energy is conserved in this respect by the fact that only the air required for adequate ventilation has to be heated. Having such precise control of the amount of air entering and leaving (since air has to be exhausted to balance the pressure), it is possible to

utilize the heat in the exhaust to preheat the intake air. Energy recovery devices like heat pipes can readily effect this exchange and their cost would probably be quickly recouped in the fuel saved. The use of such heat exchangers can reduce the winter space heating requirements by one-third; and they can reduce summer air conditioning by one-sixth.[13] However, in residences with less control of air changes, such devices would be relatively useless.

Infiltration creates two winter comfort problems. The first is heat loss and the second is dehumidification. (At 35°F and 80% relative humidity, there is only enough moisture in the air to produce approximately 20% relative humidity at 70°F.)[14]

Figure 1.8 is a graph from ASHRAE that shows the infiltration rate for various types of timber construction.[15] It is measured in cubic feet per hour and

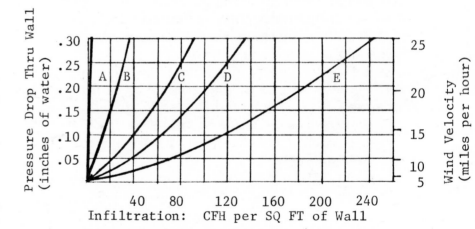

A: Same as curves B and D with building paper added.
B: 16" shingles on shiplap.
C: 24" shingles on shiplap.
D: 16" shingles on 1x4" boards on 5" centers.
E: 24" shingles on 1x6" boards on 11" centers.

Figure 1.8

is dependent on the wind speed, since the pressure difference between the inside and outside is responsible for pushing air through the wall on the windward side and sucking it out on the leeward side of a building. The amount of air moving through a wall also depends on the tightness of construction. So, even a well-insulated building can be a big energy waster if construction is poor. Curve "A" in Figure 1.8 is for the same construction types as curves "B" and "D" except that building paper has been added. The addition of a membrane, like building paper, can do a lot to cover all the cracks in timber construction, and reduce infiltration to a minimum. (Note: average winter wind velocity in the St. Louis area is a little over 10 mph.)

Figure 1.9 is an ASHRAE table that compares infiltration of brick construction to frame.[16] In brick construction, the plastering creates a membrane that reduces infiltration: it seals the cracks between the bricks. In comparison, there is little difference in infiltration between types of walls—if they are well constructed.

Type of Wall	Wind Velocity (miles per hour)					
	5	10	15	20	25	30
Brick Wall						
8.5" Plain	2.0	4.0	8.0	12.0	19.0	23.0
Plastered	0.02	0.04	0.07	0.11	0.16	0.24
13" Plain	1.0	4.0	7.0	12.0	16.0	21.0
Plastered	0.01	0.01	0.03	0.04	0.07	0.10
Frame Wall						
Lath and Plaster	0.0	0.07	0.13	0.18	0.23	0.26

Cubic Feet / Square Feet-Hour

Infiltration Through Walls

Figure 1.9

Figure 1.10 shows the importance first of well-fitted windows; and, secondly, of good weather-stripping.[17] Infiltration for windows (and doors) is a function of the length of cracks around openings. So,

Type of Window	Comments	Wind Velocity (mph)					
		5	10	15	20	25	30
Double Hung Wood Sash (Unlocked)	Total for average window, non-weatherstripped, 1/16" crack and 3/64" clearance. (Includes leakage around frame.)	7	21	39	59	80	104
	Same as above, weatherstripped.	4	13	24	36	49	63
	Total for poorly fitted window, non-weatherstripped, 3/32" crack and 3/32" clearance. (Includes wood frame leakage.)	27	69	111	154	199	249
	Same as above, weatherstripped.	6	19	34	51	71	92
Double Hung Metal Windows	Non-weatherstripped, locked	20	45	70	96	125	154
	Non-weatherstripped, unlocked	20	47	74	104	137	170
	Weatherstripped, unlocked	6	19	32	46	60	76

Infiltration Through Windows

Figure 1.10

immediately, the type of window has a direct impact on infiltration. If windows with the same shape and size are compared it is obvious that fixed glass (Figure 1.11) has the least perimeter cracks (none). Single hung (Figure 1.12) is next, followed by casement (Figure 1.13) and double hung (Figure 1.14). The cracks in awning (Figure 1.15) and hopper (Figure 1.16) type windows depend on the number of operating panels—number of panels times the perimeter of each panel. Although there is little difference between the poorly fitted wooden frame and the metal one, it should be understood that this is for infiltration only; heat loss by conduction will be higher through the metal frames than through the wooden ones. Since the metal frames lose more heat, it is suggested that wooden frames be used whenever possible.

The values in Figure 1.10 include the crack created by the window frame. The values for the window frame alone are tabulated in Figure 1.17.[18] There is little that weatherstripping alone can do to effect reductions in these cracks around the frame. However, properly installed storm windows can help somewhat if they are well sealed and cover the entire frame opening (Figure 1.18A). If the frame cannot

| | Wind Velocity (mph) | | | | | |
	5	10	15	20	25	30
Masonry wall -- uncaulked	3	8	14	20	27	35
-- caulked	1	2	3	4	5	6
Wood frame construction	2	6	11	17	23	30

Infiltration Around Frame

Figure 1.17

be covered, it needs to be well caulked where it meets the wall. Too often, storm windows are loosely fitted inside the window frame (Figure 1.18B) where the only good they do is to cut down on heat loss through the glass.

Windows which have to slide in the sash (double and single hung and sliding windows) need to be relatively loose fitting in order to operate properly. This means that they have to be fitted so that they move freely in the spring and summer when windows are most used and when the expansion of the wooden window and its frame is greatest due to ther-

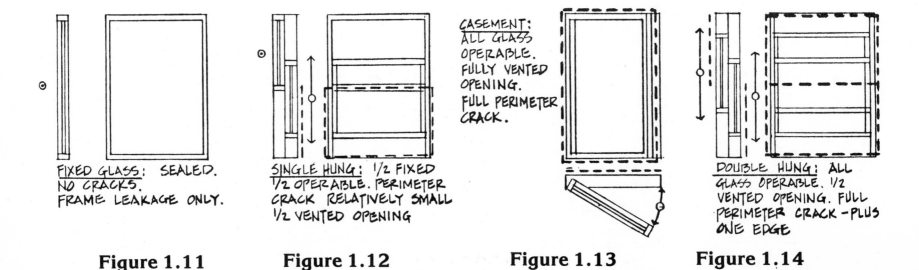

FIXED GLASS: SEALED. NO CRACKS. FRAME LEAKAGE ONLY.

SINGLE HUNG: 1/2 FIXED 1/2 OPERABLE. PERIMETER CRACK RELATIVELY SMALL 1/2 VENTED OPENING

CASEMENT: ALL GLASS OPERABLE. FULLY VENTED OPENING. FULL PERIMETER CRACK.

DOUBLE HUNG: ALL GLASS OPERABLE. 1/2 VENTED OPENING. FULL PERIMETER CRACK - PLUS ONE EDGE

Figure 1.11   **Figure 1.12**   **Figure 1.13**   **Figure 1.14**

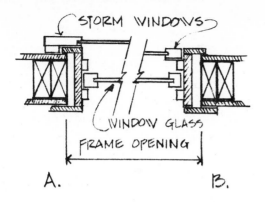

A.　　　　　　　B.

Figure 1.18

Thermal Expansion per °F

	Parallel[18] to Fiber	Across[19] Fiber
Fir	2.1×10^{-6}	21.0×10^{-6}
Maple	3.6	36.0
Oak	2.7	27.0
Pine	3.0	30.0

(inches per inch)

Figure 1.19

mal expansion across the grain (Figure 1.19) and the swelling caused by the relatively higher moisture content of the air. Since they have to fit loosely in their track during their most swollen period, they will fit even more loosely during the winter when thermal expansion becomes contraction, and the moisture content of the air is at its lowest. Since infiltration is a function of pressure difference, weatherstripping—to be most effective—has to resist this pressure dif-

ferential. This is fairly easy to do at the top and bottom of double hung windows, since the locking mechanism can force the window tightly against felt strips on the frame. But this pressure resistant seal is hard to achieve along the side track of the window. On the other hand, casement type windows, which swing open, can be locked down against spring brass or aluminum strips in the sash (Figure 1.20). This gives a pressure-tight seal that adjusts to the variances of the window as it expands and contracts. For such windows, it is less important to hang a storm window for control of infiltration. This means that if double insulating glass is used in a casement window that has good pressure-resistant weatherstripping, it

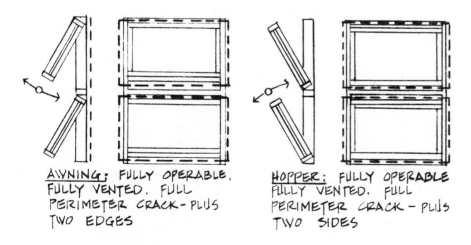

AWNING: FULLY OPERABLE. FULLY VENTED. FULL PERIMETER CRACK - PLUS TWO EDGES

HOPPER: FULLY OPERABLE FULLY VENTED. FULL PERIMETER CRACK - PLUS TWO SIDES

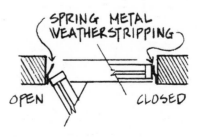

Figure 1.15　　　　**Figure 1.16**　　　　**Figure 1.20**

can probably get by without storm windows. This would allow much greater freedom to take advantage of the natural ventilation of those days that are pleasant in the fall, spring and summer that would otherwise be sealed out by the storm windows, which are too cumbersome to open or to take down for just one or two days. (Note: storm windows and weather-stripping are still important in the summer to keep hot air out, if air conditioning is required.) This freedom to take advantage of naturally good days will mean greater energy savings simply because all the machines can be turned off. This is one of the reasons casement windows might be more favorable than the currently popular hung windows. (The same pressure-type mechanism can also be used with awning and hopper type windows.)

In dealing with doors and the problem of infiltration, the best solution is to design the house such that doors open into buffer spaces. If kitchen doors open into garages or mud rooms, or if front doors open into vestibules or enclosed porches, they provide a series of pressure reducing porches. This creates successive pressure drops that reduce the force of infiltration. In addition, these unheated, but enclosed, buffer spaces also reduce heat loss through conduction, since they are typically at a temperature half-way between inside and outside temperatures. These air-locked type entries can make a large difference in heat loss especially if doors have to be located in the wall that faces the prevailing winter winds.

CEILINGS AND ATTICS

Since heat rises, ceilings and roofs generally have marginally greater thermal pressure on them to transmit heat out of the building during the winter. During the summer, roofs receive the brunt of the solar radiation and tend to transmit this heat into the dwelling. For these reasons it is important to put extra insulation in the ceiling in order to make sure it does not lose or gain heat at a faster rate than the walls.

If the house has a pitched (rather than flat) roof, the attic space is a natural thermal buffer. During the winter, the attic without insulation should be as tightly sealed as possible to better hold the heat transmitted through the ceiling. In an insulated attic vents are required to allow moisture to escape before it can condense. During the summer, all attics should be well ventilated to provide cooling for the roof. The temperature of the attic in winter will be somewhere in between the temperature inside the dwelling and the temperature outside; and the heat transmission can be adjusted downward because of the lower temperature difference.

If the roof is flat, then the heat loss would be calculated as for a wall, with one modification: the interior temperature can be adjusted up about 5°F to accommodate the natural stratification of temperatures in the room. Insulation in the roof should be beefed up to reduce transmission to approximately the same rate as that for the walls to ensure that the effort to reduce heat transmission is not disproportionately applied.

FLOORS

If floors are over crawl spaces, extra effort should be taken to seal the floor and insulate it; and, like the attic in winter, the crawl space should be closed up so that it can be a buffer space below the floor. Its temperature will be somewhere between the outside and inside temperatures.

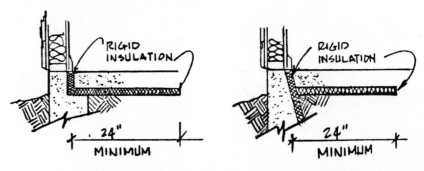

Figure 1.21

If the floor is on a slab on grade, heat loss to the ground immediately beneath the slab is negligible. Heat loss due to the slab is a function of the proximity of the slab **edge** to outdoor conditions. If the edge of the slab is not insulated, the floor inside the house will be cold and uncomfortable. It is strongly advised that the slab be insulated with 24" of rigid, moisture-resistant insulation. Insulation for the slab has to be placed to cut off any direct path of conduction. Figure 1.21 shows several suggested slab/insulation arrangements. The insulation has to be placed at least over the edge of the slab; but pulling the insulation 24" back underneath the slab reduces heat losses through ground conductance. Heat is often introduced at slab edges because, even with insulation, the floors can still be uncomfortably cold. This requires a lot more energy; so, instead of heating the slab, it is suggested that perhaps increasing the amount of insulation used could ultimately lower costs and energy requirements. Generally 1.5" (resistance = 5.0) is the heaviest insulation used; so, if the thickness (or the R value) is doubled, heat loss would be greatly reduced and the added expense of pipes or wires embedded in the slab can be minimized.

An approach to effective analysis of the impact of various energy conserving measures is presented in Appendix E.

VENTILATION

Once a hole has been placed in a wall for a window, it only makes sense that the entire opening should be available for ventilation.

However, most recent residential construction uses aluminum, single hung windows (Figure 1.22) which are the least effective for achieving comfort through ventilation. This type of window is, however, very appropriate for the sealed, totally air conditioned residence. It will be opened only on rare occasions, so there is no reason to buy a more expensive type. This raises the question of the appropriateness of a totally air conditioned residence. The only real reason to seal a building—since comfort can be achieved through design—is for health reasons. If the occupants suffer severely from allergies, or if pollution is a serious or obnoxious problem, then controlled ventilation with adequate filtration is probably the best route. This calls for a totally air conditioned environment for which single hung windows are more than adequate.

Double hung windows (Figure 1.23) are the next level of response. They are superior to single hung because they promote the positive removal of hot air in the dwelling. As air heats up it rises. In the case of single hung windows hot air is trapped in the upper half of the room (Figure 1.22). With double hung, the window can be opened at the top, on the leeward side, to exhaust the hot air as it rises. The net effect is

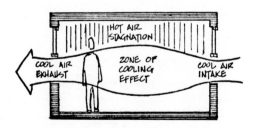

Figure 1.22

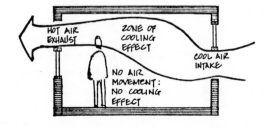

Figure 1.23

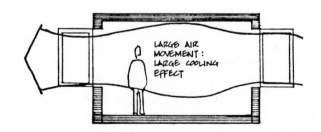

Figure 1.24

15

extra cooling. However, there is one major disadvantage to sash, or sliding windows: the glass itself obstructs the opening. Even opened to their maximum, they cut the effective area of the wall opening in half. The amount of air that can be moved is a direct function of the size of the openings. And, anything that reduces the size of the openings also reduces the cooling effect. (The cooling effect is also a function of the air velocity.)

For the same amount of glass area, casement type windows (Figure 1.24) provide about twice the opening, since they swing out of the window frame. The use of the hole in the wall that the window sits in is maximized for ventilation. The hot air will still be exhausted, and even more air can be moved through the building, so the possible cooling effect becomes even greater.

Casement windows can swing either to the inside or to the outside of the wall (Figure 1.25). If it swings out, then the screen has to be located on the inside and the screen either has to open to get to the window, or there has to be extra hardware, like a crank, to open the window without moving the screen. In addition, if the window does not swing open 180°, but stands out from the wall like a fin, it can actually deflect some of the possible ventilation on the positive, or windward, side of the building. It does not make much of a difference on the leeward side (except that the leeward can become the windward if the wind shifts).

A casement window which opens in has the screen on the outside where it can protect the glass; there is no need for cranks or screens that open (which may mean lower costs); and the windows do not impede ventilation. The disadvantage with inward swinging casement windows is that they might interfere somewhat with furniture arrangements.

Awning type windows which project outward (Figure 1.15) do not impede the air flow like outward-swinging casement windows. And, unlike all the other types, they can be left open during rainstorms because they keep the water out. This may be an im-

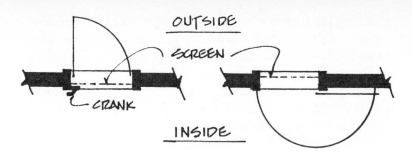

Figure 1.25

portant consideration in climates where there is a lot of rain and ventilation is needed to provide adequate cooling effect. The major drawback to the awning type is their extra expense because of all the hardware required to crank it up and lock it in position (the screen has to be on the inside).

INSULATION

Since insulation is generally needed at all surfaces, it can become a significant cost item in construction. It is important to carefully consider the amount of insulation needed. One needs to find out at which point the incremental effect of added insulation ceases to show sufficient returns in less equipment and lower fuel requirements.

In general, the best way to insulate masonry walls (in terms of time and materials at least, if not cost as well) is to affix rigid sheets of insulation (like expanded polystyrene) to the surface of the wall.

In frame construction the depth of the stud generally determines the amount of insulation that can be placed in the wall. Recommendations for minimum wall resistance to heat transfer, however, are beginning to reach the point where this thickness (3.5") will be insufficient if the standard mineral wool or fiberglass insulating bats are used. To compound the problem, the suitability of some insulating foams that were supposed to meet the need for high resistivity is in doubt because of some of their flame spread characteristics, or the fact that they give off toxic gases in the event of a fire. Foams will probably be developed that meet the requirements of high resistance and safety, so the architect should keep abreast of this particular industry.

However, bats and foams achieve their overall average resistivity by filling the cavity between studs. As insulating techniques have gotten more sophisticated it has been noted that the studs can represent 12% of the wall area; and that these studs represent a solid path of conduction for heat loss straight through the wall from the inside to the outside in winter (vice versa in the summer). As the transmission through the rest of the wall gets lower, heat transmission through studs becomes more and more significant until most of the heat is moving through that 12% of the wall. To more effectively retard heat loss, there are rigid foam boards that can be used like exterior sheathing.[21] This places a layer of insulation **over** the studs and effectively severs this conducton path. Another fact that has come out of further studies on insulation, is that insulation is more effective if it is placed on the exterior of the building.[22] Placing insulation on the outside makes sure that it intercepts those paths of conduction similar to that with the studs, and it also places the insulation at the point of maximum temperature difference where its resistance will be utilized to its fullest. For example, if the insulation is in the wall, it is being insulated by the sheathing and the siding, which lowers the temperature difference across the insulation. In response to this, insulation is available that has an exterior, finished surface; it just has to be mounted on the base wall, either frame or masonry, and the job of insulating and finishing are completed in one step.

EQUIPMENT

Another area where significant energy savings can be reaped is in the size of the heating and cooling units put in residences. Generally, the equipment a builder puts in a home is oversized. The builder has to try to please a public that demands precise temperature control; and his capability to perform an adequate heat loss and heat gain analysis is limited.

So, if the cost is to be passed on to the buyer, then it is better to play it safe—besides, a little extra never hurts.

This attitude and response yields home heating and cooling systems that operate very wastefully. The efficiency of heating and air conditioning is especially sensitive to the percentage of its full working capacity that is being used. The equipment is designed to operate most efficiently when it is working under a full load. Under full load conditions new furnaces convert the fuel used to useable heat at 75% efficiency. Poorly maintained furnaces drop to efficiences as low as 35%.[23]

Obviously, it is important to maintain the equipment and try to run it very near to its rated capacity to make the best use of the fuel. But, the usual practice is to design heating and cooling systems (when an actual analysis is run) based on outdoor conditions that are only exceeded 2.5% of the time. This means that the system is working at its maximum very infrequently. The rest of the time it is working under partial loads at lower efficiencies.

It has recently been suggested that, in order to conserve energy, the designs should be based on outdoor conditions that are exceeded 50% of the time.[24] This lowers the required capacity and slightly undersizes the system compared to current practice. The result is that the system operates more frequently at its rated capacity; and therefore, more efficiently. In terms of thermal comfort, the system will "fail" to produce its indoor design temperature about 5% of the time. This represents a few hours a year (less than 30) when the spaces will be warmer or cooler than the present "standards."[25] This seems a small price to pay for better energy conservation.

Another cause of energy wastefulness is setting a very narrow temperature range on thermostats. The equipment has to come on very often for only short periods. It brings the temperature up—say 2°F—and cuts off. This constant cycling on and off for such small loads is very costly in terms of energy. The energy is much more effectively used when operating with large swing temperatures. The greater the tolerance one has for a range of temperatures, then the greater will be the energy savings.

It is important to decide what comfort levels are required and have a thorough heat loss/heat gain analysis done to ensure that the equipment is properly sized so that it will not be wasting energy.

Some of the above techniques for energy conservation may increase the costs in some areas but it is expected that most, if not all, of these increments in cost will be compensated for in the savings realized when much smaller heating and cooling units are sufficient. Harold Hay (See Chapter 3) had to spend more for roof ponds and moveable insulation, but he was able to completely eliminate the need for auxiliary space conditioning systems.

The effective utilization of energy and its real place in energy conservation, and the fact that other scales beside residences and office buildings are also targets for energy conservation are illustrated by a remarkable school built in 1961 in England.[26] The school is St. George's Secondary School in Wallesly near Liverpool, and it was designed by Emslie Morgan. The main features are: it is masonry construction throughout for large thermal mass; it has 5" of polystyrene foam over the roof slab and on the north side of the building; and it has a double-skinned, south-facing glass wall that turns the whole building into a collector. The two skins of glass are separated by a .6m (24") air space. The mass of the building that is right behind the inner skin is painted black. The inner skin is primarily obscured glass so that it passes diffused light into the classrooms. In some places the inner skin is clear glass and it is backed by reversable aluminum panels that are black on one side for winter absorption and white on the other for summer heat reflection.

The building takes advantage of all sources of heat. The heat from the lights is used and the body warmth of occupants even accounts for one-sixth of the heating supply. The main portion, however, is solar; and it is pumped into the building and held in its mass. This thermal mass helps keep the temperatures relatively warm throughout the night, and in the morning the lights and body heat help bring the temperature into the comfort range. In this way, baseline heat is provided by the sun for the 24 hour period, but when the building is in use, the two items directly associated with its being used—light and people—supply any extra heat needed. When the building is not in use, no additional energy is required. The energy that is naturally available when the building is in use is so well utilized that an auxiliary heating system that was installed, because officials doubted whether the solar design would work, has been removed because it was never used. In the murky north of England, at 53° North Latitude, this school simply does not need an auxiliary system for space heating; in fact, it often overheats, even when temperatures are as low as 20°F. It seems that similar designs could accomplish at least as much in the more southerly climates and generally clearer weather of the United States of America. It can be done, all that lacks is the doing.

Energy conservation is the first step forward to energy self-sufficiency. Real conservation will come when attitudes change and lifestyles begin to reflect a more sensible, less wasteful utilization of what energy is available.

NOTES

1. C. Steinhart, and J. Steinhart, **Energy's Sources, Use and Role in Human Affairs** (North Scituate, Mass.: Duxbury Press, 1974), p. 209.

2. F. S. Dubin, "Energy Conservation through Building Design and a Wiser Use of Electricity," Paper at the Annual Conference of the American Public Power Association (San Francisco, California), p. 2.

3. J. R. Wright and P. R. Achenbach, **Scientific American Roundtable on Energy Conservation in Buildings** (August 29, 1973), p.8. Comments of Dr. Betsy Amber-Johnson, Asst. Secretary for Science and Technology, U.S. Department of Commerce.

4. **Ibid**., p. 5.

5. **Ibid**., p. 9. Comments of Arthur F. Sampson, Administrator, General Services Administration.

6. Ken Labs, The Architectural Use of Underground Space. Issues and Applications (Washington University, St. Louis, Mo.: Masters of Architecture Thesis, May, 1975), p. 36.

7. A. L. Hammond, "Individual Self-Sufficiency in Energy," **Science**, Vol. 184, (April 19, 1974), p. 281.

8. F. W. Hutchinson, "The Solar House: Analysis and Research," **Progressive Architecture** (May 1947), p.90-94. Although large south glass areas result in more collected solar energy, the structure of the house has to be able to absorb the extra heat to keep the south rooms comfortable, and it has to be able to release the heat later when it is needed. With frame construction (low thermal mass), rooms with south glass become way overheated and (1) the heat has to be exhausted or (2) drapes, or the like, are used to block out the sun. The south glass may gain enough heat to offset the loss of energy through the extra glass at night, but the energy may be unusable because the structure cannot modulate the way the energy is used.

9. International Conference of Building Officials, **Uniform Building Code**, 1973 ed., p. 88.

10. W. J. McGuiness and B. Stein, **Mechanical and Electrical Equipment for Buildings**, 5th edition (New York: John Wiley and Sons, Inc., 1971), p. 169.

11. Charles A. Berg, "Energy Conservation through Effective Utilization," **Science**, Vol. 181 (July 13, 1973), p. 130.

    G. E. Mattingly and E. F. Peters, "Wind and Trees—Air Infiltration Effects on Energy in Housing," (Center for Environmental Studies, Report No. 20, May, 1975, Princeton University, NSF Grant No. 11237), p. 6.
    This report suggests that infiltration heat losses may be as much as 75% of the total heat loss—especially if exterior doors are open from 1-2 minutes and hour.

12. W. J. McGuiness and B. Stein, **op. cit.**, p. 171.

13. Philip Steadman, **Energy, Environment, and Building** (Cambridge: Cambridge University Press, 1975), p. 34.

14. W. J. McGuiness and B. Stein, **op. cit.**, p. 238.

15. ASHRAE, **Guide and Data Book: Fundamentals and Equipment** (New York: ASHRAE, 1966), p. 460.

16. **Ibid.**, p. 460.

17. **Ibid.**, p. 459.

18. **Ibid.**, p. 459.

19. J. N. Boaz, editor, **Architectural Graphic Standards,** 6th edition (New York: John Wiley and Sons, 1970), p. 313.

20. H. D. Tiemann, **Wood Technology** (New York: Pitman Publishing Co., 1942), p. 234.

    Tiemann suggests that cross-fiber thermal expansion is ten times the longitudinal, expansion rate.

21. **Sweet's Architectural Catalogs** (New York: McGraw Hill, 1971), Section 7.14, p. 18.

22. General Services Administration, **Energy Conservation Guidelines for Office Buildings** Dublin: Mindell Bloom Associates, Jan., 1974), pp. 9-14.

23. C. A. Berg, **op. cit.**, pp. 130-131.

24. F. S. Dubin, **op. cit.**, p. 6.

25. **Ibid.**, p. 6.

26. P. Steadman, **op. cit.**, pp. 35-36, 160-161.

CHAPTER 2

THE SUN WITH THE EARTH

The sun is a hydrogen-fusion, nuclear furnace that is powered by sheer gravitational crush. The mass of the sun is 1.97×10^{30}kg (2.17×10^{27}tons). Although the radius of the sun is almost three times the distance between the earth and the moon, the gravitational pull at the surface of the sun is 28 times the gravitational pull at the surface of the earth. (The pull of gravity decreases proportionally with the second power of the distances between the centers of attracting bodies: i.e., doubling the distance cuts the pull to ¼ of its former force.)

To fuel this furnace, matter is consumed at the rate of 4 million tons a second. Even at this rate, the sun is expected to last another 4.5 BILLION years. As this fuel is consumed, some of this matter is converted to light energy and it is radiated away from the sun. This light expands through space in spherical waves. In the eight minutes it takes the light to travel the 93 million miles to earth, the spheres of light have expanded to such an extent that the 1.27×10^8km² projected area of the earth intercepts only 5 billionths (5×10^{-9}) of the energy radiated from the sun. Because of the tremendous distances involved, fluctuations of energy output at the surface of the sun become so attenuated that the solar energy reaching the outer atmosphere can be considered constant.[1]

Currently, this constant is defined as 2.00 gram calories per square centimeter per minute (2.00 gm cal/cm²·min) — ± 2%.[2] The gram, or small, calorie is the designation used to differentiate it from a kilocalorie (1000 calories) which is often called a calorie when referring to the energy content of food. In the terminology of solar meterological data, **1.0 gm cal/cm²** is called a **langley**. So the solar constant is also 2.0 langleys per minute (442 Btu/ft²·H). This is the amount of energy that would fall on the earth if it were a flat surface held perpendicular to the sun's rays at the average distance to the sun and if the earth had no atmosphere. At the earth's surface, this energy will always be less than 2.0 langleys per minute.

However, the earth does move, it is a sphere and it

SUN DIAMETER: 1.39×10^6 KM
8.65×10^5 MILES

GRAVITY AT SUN'S SURFACE: 28 G's

SUN RADIUS: 6.95×10^5 KM
4.32×10^5 MILES

SOLAR MASS: 1.97×10^{30} KG
2.17×10^{27} TONS

RELATIVE EARTH TO MOON DISTANCE: 3.84×10^5 KM
2.38×10^5 MILES

RELATIVE SIZE OF THE EARTH TO THE SUN.
EARTH DIAMETER: 1.27×10^4 KM
7.92×10^3 MILES

has an atmosphere. These factors are responsible for the fact that there is a wide range of flux over the surface of the earth at any one moment. Understanding these variances is important to the wise utilization of solar energy.

The earth moves. It moves with three distinct motions:
1. It moves around the sun.
2. It moves toward and away from the sun.
3. It spins on its axis.

The motion of the earth with respect to the sun is made even more complex by the fact that the axis of the earth is tilted 23°27' (almost 23.5°) from a line perpendicular to the plane of the earth's orbit around the sun. (It can also be stated that the axis is tilted 66.5° from the orbital plane, but convention expresses it as 23.5° from perpendicular to the orbital plane.)

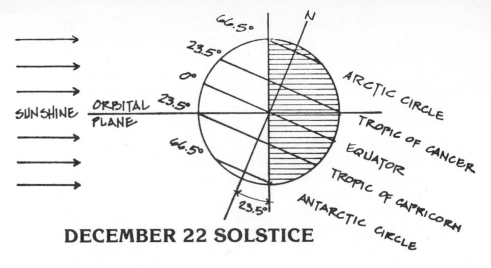

DECEMBER 22 SOLSTICE

Figure 2.3

As the earth circles the sun, the axis of the earth sweeps out a skewed cylinder. Twice a year, at the equinoxes, the axis is coplanar with a cylinder that is perpendicular to the earth's orbital plane and parallel to the axis of the sun (Figure 2.2).

At the spring and fall equinoxes (March 22 and September 22), the north and south poles lie in a plane perpendicular to the sun. The significance of this is that there are equal hours of day and night over the entire sphere of the earth. The length of day and night progressively increase or decrease as the earth moves toward the winter and summer solstices. In the northern hemisphere, the days get shorter and the nights longer as the December 22 solstice approaches. (In the southern hemisphere, the days get longer and the nights shorter.) At this solstice, the north pole is tilted its maximum of 23.5° away from the sun. As a result, all points farther north than 66.5° N. Latitude have a 24 hour night, and all points farther south than 66.5° S. Latitude have a 24 hour day because the south pole is tilted its maximum of 23.5° toward the sun (Figure 2.3). The half-shell of solar illumination shifts over the surface of

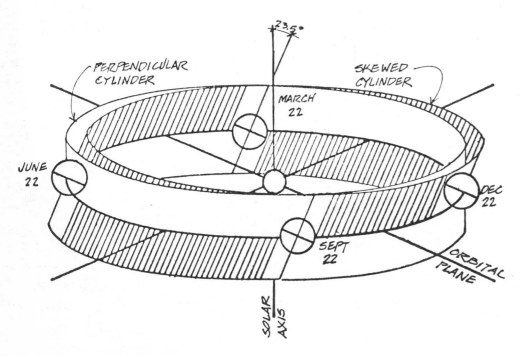

Figure 2.2

the earth as it goes through the equinox toward the June 22 solstice (Figure 2.4). There the situations are reversed, and the north pole becomes inclined 23.5° toward the sun. The tilt of the earth is responsible for the seasons that occur on the earth.

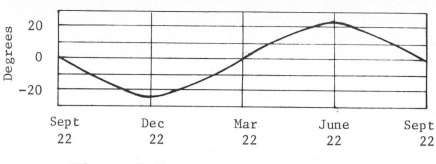

Figure 2.5

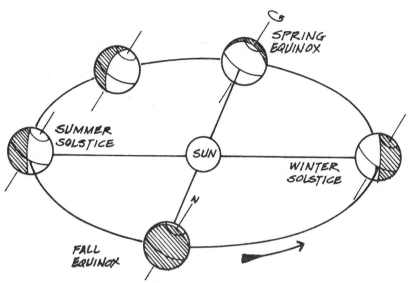

Figure 2.4

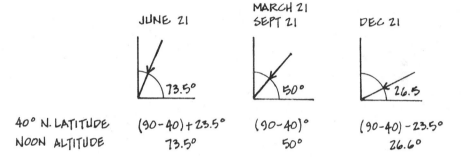

40° N. LATITUDE	$(90-40)+23.5°$	$(90-40)°$	$(90-40)-23.5°$
NOON ALTITUDE	73.5°	50°	26.6°

Figure 2.6

As the seasons change, the position of the sun in the sky, at noon, changes relative to the surface of the earth. (It changes for all other hours as well, but noon is the reference angle.) Figure 2.5 is a chart that plots the declination angle of the sun as a function of the time of year. The sun appears to move through an arc of 47°. This arc is divided into two segments of 23.5°: this establishes the equinoxes as the reference angle of 0° deviation. For any latitude, the angle of the sun above the horizon (solar altitude) at the equinoxes is equal to 90° - Latitude° (Figure 2.6). For the solstices, the altitude at noon is equal to (90° - Latitude°) ± 23.5°.

The earth travels around the sun in an almost perfect circle, but the sun is not at the center of this circle. As a result, the earth is closest to the sun around January 1; and it is farthest away around July 1. There is a difference of more than 3 million miles between these extremes. This is a difference of 3.3% and, since the intensity of solar radiation varies inversely with the square of the distance between the earth and the sun, the solar energy reaching the earth is almost 7% stronger in January than in July (Figure 2.7).[3]

Because the earth spins on its axis, the half-shell of solar illumination travels over most of the sphere at periodic intervals. This produces night and day. The spin ensures that there is even distribution of the solar energy that is available. The fluctuation of energy at any one place—over time—is a function of the tilt of the earth and its accompanying variation in the length of day (Figure 2.11).

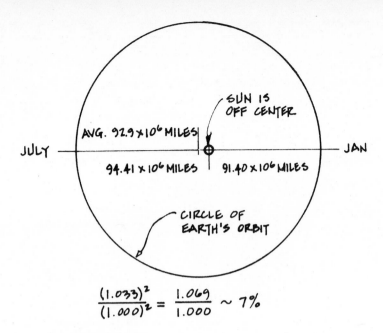

$$\frac{(1.033)^2}{(1.000)^2} = \frac{1.069}{1.000} \sim 7\%$$

Figure 2.7

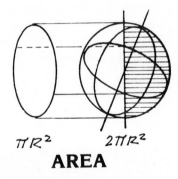

AREA

Figure 2.8

The earth is basically a sphere; and, because it is a sphere its surface is twice as large as its projected area. Figure 2.8 shows the relation between these two surfaces. Light that passes through the projected area of the earth falls on increasingly larger areas as the surface of the globe curves away from the sun. This dilutes the power per unit area. Figure 2.9

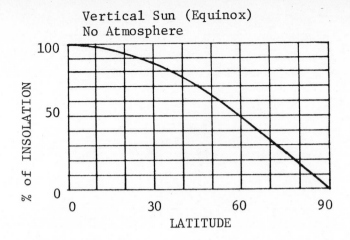

Vertical Sun (Equinox)
No Atmosphere

[4]**Figure 2.9**

shows how this energy varies as the surface curves away from the perpendicular.

Because the earth is a sphere that tilts and turns, the length of day varies quite a lot as the latitude becomes greater and the axis either inclines toward or away from the sun. Sections (Figure 2.10) at in-

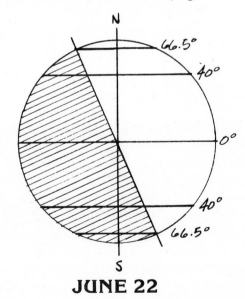

JUNE 22

Figure 2.10

Length of day in various northern latitudes (In hours and minutes on the 15th of each month)

Month	0°	10°	20°	30°	40°	50°	60°	70°	80°	90°
Jan.	12:07	11:35	11:02	10:24	9:37	8:30	6:38	0:00	0:00	0:00
Feb.	12:07	11:49	11:21	11:10	10:42	10:07	9:11	7:20	0:00	0:00
Mar.	12:07	12:04	12:00	11:57	11:53	11:48	11:41	11:28	10:52	0:00
Apr.	12:07	12:21	12:36	12:53	13:14	13:44	14:31	16:06	24:00	24:00
May	12:07	12:34	13:04	13:38	14:22	15:22	17:04	22:13	24:00	24:00
June	12:07	12:42	13:20	14:04	15:00	16:21	18:49	24:00	24:00	24:00
July	12:07	12:40	13:16	13:56	14:49	15:38	17:31	24:00	24:00	24:00
Aug.	12:07	12:28	12:50	13:16	13:48	14:33	15:46	18:26	24:00	24:00
Sept.	12:07	12:12	12:17	12:23	12:31	12:42	13:00	13:34	15:16	24:00
Oct.	12:07	11:55	11:42	11:28	11:10	10:47	10:11	9:03	5:10	0:00
Nov.	12:07	11:40	11:12	10:40	10:01	9:06	7:37	3:06	0:00	0:00
Dec.	12:07	11:32	10:56	10:14	9:20	8:05	5:54	0:00	0:00	0:00

[5]Figure 2.11

creasing latitudes show that as the circles get smaller, greater percentages of these circles fall within the half-shell of illumination during the summer. These percentages reflect how much of the 24 hour period is day light. This accounts for the midnight sun during the summer and, conversely, the long winter nights when the axis is pointed away from the sun. Figure 2.11 is a table of the hours and minutes of sunlight for different latitudes each month of the year (assuming the 15th day of the month will yield the average length of day).

Figure 2.12 presents the maximum and minimum (summer and winter) daily insolation values for different latitudes. Figure 2.13 shows the total, annual solar radiation for different latitudes. Figure 2.14 presents the annual cycle of varying energy at select latitudes. The combination of these figures brings out a little known fact: the highest daily and annual insolation levels do not occur at the equator; they occur around the latitudes of the Tropics of Cancer and Capricorn. This is a result of the earth's tilt and its curvature. At the solstices, latitudes like 40° are more perpendicular to the sun's energy than the equator. In addition, as one gets closer to the poles, the energy becomes more dilute; but the number of hours this energy is available increases dramatically.

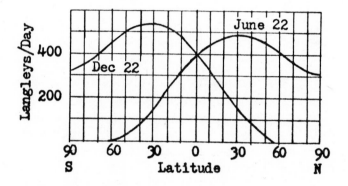

[6]Figure 2.12

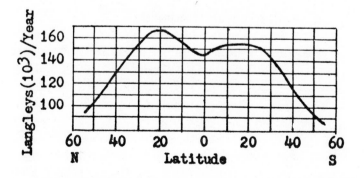

[7]Figure 2.13

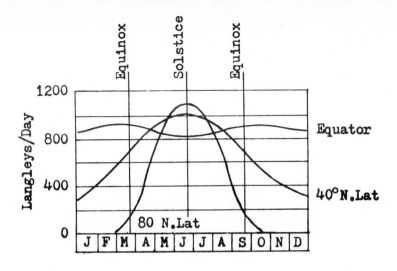

[8]**Figure 2.14**

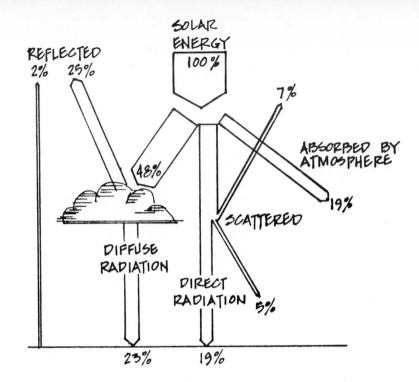

[9]**Figure 2.15**

The poles receive only slightly less than the equator at the solstices, but it requires twice the time. According to Figure 2.14 the equator should receive the greatest yearly total of energy, since its minima are not much less than its maxima. (Note that the equator has "Summer" twice a year—at the equinoxes.) Figure 2.14, however, is based on extra-terrestial insolation and this neglects the effect of the atmosphere. Figure 2.13 is based on observed, direct and diffuse insolation at the earth's surface; and the dip in energy at the equator reflects the effect of typical, tropical cloudiness.

The earth's atmosphere is responsible for most of the reduction in solar energy reaching the ground. It is also responsible for the fact that solar energy reaches the earth by two different mechanisms: direct and diffuse radiation. This all results from some rather complex interactions between sunlight and the atmosphere (Fig. 2.15). A considerable portion of this solar energy is reflected back into outer space by the atmosphere and the tops of the clouds. Some of the energy is absorbed by actual air molecules, particularly by oxygen, ozone, carbon

dioxide and water vapor. In addition, gas molecules, dust particles, pollutants and water droplets scatter the light more or less uniformly in all directions. This scattering mechanism affects short wavelength (blue end of the spectrum) light the most. As a result, the sky appears uniformly blue. Water droplets scatter light very strongly; and, where there are dense aggregations of water droplets (heavy clouds), as much as 80% of the incident light is scattered back to space. The average cloud cover for the earth is about 50%, which makes scattering a mechanism for substantial loss of solar energy.[10]

Oxygen and ozone molecules absorb high energy, ultraviolet light to fuel certain chemical reactions. Carbon dioxide and water vapor absorb certain wavelengths of light that resonate with the frequency of their molecular vibrations. This mechanical energy is then distributed through the atmosphere by collisions with other air molecules. As the energy absorbed by carbon dioxide and water vapor is redistributed, it is reradiated. The atmosphere, itself, becomes a source of stored solar energy. This is why there is a considerable amount of energy coming through on even the most overcast days (300 W/m² on a horizontal surface), and why temperatures do not drop drastically during the night. Even in arid regions, where there is little cloud cover to hold and reradiate this energy, the intensity only rarely drops as low as 100 W/m².[11] Without this atmospheric radiation, the surface temperature of the earth would drop radically at night through radiation to outer space. Fig. 2.16 shows the distribution of solar energy by wave-length outside the earth and at the earth's surface. The general downward shift is due to reflection and light scattering of the atmosphere. The large dips in the curve correspond to those wave-lengths of light that are absorbed by molecules in the atmosphere. Fig. 2.17 is a heat balance diagram that illustrates the effect of the atmosphere in distributing heat.

According to Byers, there are zonal differences in this heat balance across the earth. The half of the earth that lies between 30°N and 30°S latitudes gains

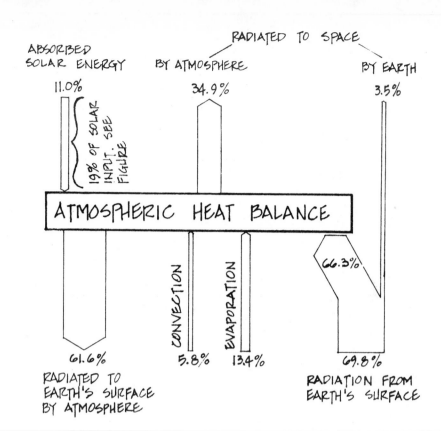

[13]**Figure 2.17**

heat by atmospheric radiation, with the atmosphere, particularly the water vapor, exerting a greenhouse effect. However, beyond latitudes 30°, N and S, the earth experiences a net loss of heat by radiation to space. For these reasons, the atmosphere and the oceans have to circulate the heat from the equator to the poles, otherwise the tropics would become progressively hotter and the higher-latitude areas would become progressively cooler.[14]

On a clear day, by the time solar radiation reaches the earth's surface, it is a little more than half the amount of energy that fell on the outer atmosphere. On heavily overcast days it can be virtually nil. Another effect the atmosphere has in reducing radia-tion is the actual length of the path of sunlight

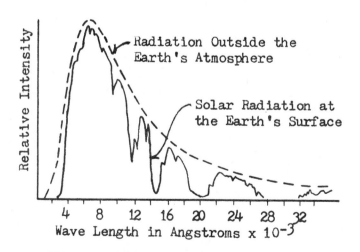

[12]**Figure 2.16**

29

through the atmosphere. Fig. 2.18 shows that as sun angles get lower, the light has to penetrate thicker sections of atmosphere.

According to Trewartha, a clear atmosphere with the sun vertically overhead transmits 78% of the solar radiation to the earth's surface. In meteorological terms this is an air mass of 1.0. Fig. 2.19 is a table from Trewartha's **An Introduction to Climate**; it shows solar altitude angles and their relative air masses. It also relates them to the amount of energy that reaches both a plane perpendicular to the sun's angle of incidence, and one that is horizontal to the ground. Fig. 2.19 shows the compound effect of reduction in energy from both the effect of the earth's curvature and the effect of the path length through a clear atmosphere.

In addition, the amount of water vapor in the atmosphere varies with the air temperature. This change in water content has quite an influence on the amount of radiation getting through. Water vapor is the controlling agent in an atmospheric absorbtion; it absorbs six times as much radiation as all other gases

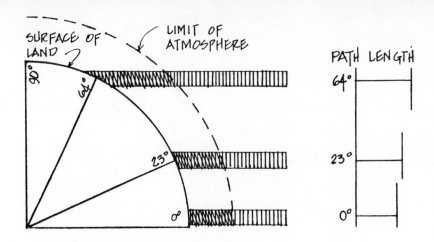

[15]**Figure 2.18**

combined. The amount of water vapor in the atmosphere varies considerably by regions and by seasons.

Because of these phenomena, regions such as the Great Lakes in North America, Central Japan and Southern France have yearly insolation totals on a par with the Amazon River region in South America.[17]

Intensity of solar radiation (Transmission coefficent 78 percent)

Sun's altitude	Distances rays must travel through atmosphere *	Radiation intensity on a surface perpendicular to rays	Radiation intensity on a horizontal surface
90°	1.00	78	78
80°	1.02	77	76
70°	1.06	76	72
60°	1.15	75	65
50°	1.31	72	55
40°	1.56	68	44
30°	2.00	62	31
20°	2.92	51	17
10°	5.70	31	5
5°	10.80	15	1
0°	45.00	0	0

* Expressed in atmospheres.

[16]**Figure 2.19**

At a smaller scale, Fig. 2.20 presents the insolation on a plane perpendicular to the sun for solar noon on the 21st of each month at 40°N latitude. This is developed from ASHRAE data for clear days. From Figure 2.6, it would be expected that December would have the least radiation because it has the lowest sun angle. However, the winter months have very low atmospheric moisture levels to interfere with insolation. The longer hours of summer sunshine are able to pump more moisture into the atmosphere and the higher temperatures are able to hold more. The net effect is that solar radiation is more intense during the winter, but that daily totals are less because of the shortened day. Note that the 7% increase in radiation during winter in the Northern Hemisphere would not compensate for the effect of air mass expected from Figure 2.19. Thus, the rest of the difference is ascribed to reduced amounts of atmospheric moisture.

A little more than half the energy intercepted by the earth reaches its surface, and this energy is spread out over an area that is twice as large as the intercepting plane. As a result, the 2.0 g.cal/cm²·min. intercepted becomes an average of 0.5 langleys/min. at the earth's surface. This energy gets through as either direct or diffuse radiation. How much of each kind reaches the surface at any one time is a function of the length of the solar path through the atmosphere and the transparency of the atmosphere. The particular mixture will vary widely for any one place. For example, it will average about 25% diffuse at Stockholm, Sweden in May, but it will be closer to 80-90% during the winter. Farrington Daniels states that in "much of Europe about half the total radiation is diffuse, with a higher proportion in winter and less in summer. At a station in Massachusetts 40 percent of the total annual solar radiation is diffuse, and in South Africa in relatively sunny climates the diffuse radiation is 30 percent of the total."[19] Liu and Jordan suggest that approximately 40 percent of radiation received at the earth's surface

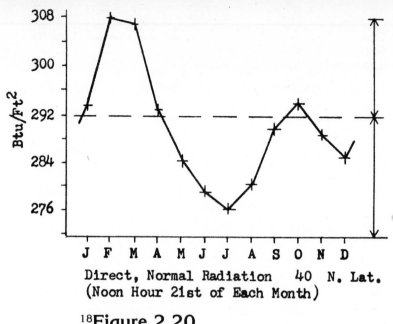

Direct, Normal Radiation   40 N. Lat.
(Noon Hour 21st of Each Month)

[18]**Figure 2.20**

is diffuse radiation.[20] Trewartha suggests that between 25 and 30 percent of total solar energy transmitted to the earth is diffuse. The fact that at least 25 percent of the expected, incoming solar energy is diffuse should be an important consideration for anyone contemplating using solar energy.

Devices used for the collection of solar energy are generally broken down into two basic types: concentrating collectors and flat plate collectors. The general nature of concentrating collectors is that they only effectively use the direct radiation component of solar energy because they require parallel light for the geometry involved in concentrating the solar energy. Diffuse light is unusable with concentrating collectors because it comes from all directions almost uniformly. Flat plate collectors, on the other hand, can utilize both direct and diffuse light without any problems. Concentrating collectors have their greatest applicability in the production of high temperatures. If high temperatures are definitely needed, then the diffuse portion of energy is lost. If the particular end use of the collected energy can be designed for lower temperatures (generally below the boiling point of water), then it is strongly suggested that flat plate collectors be used in order to take advantage of all the available solar energy.

In summary, it can be said that changes in the relative availability of solar energy due to the spin, tilt and orbit of the earth are all predictable phenomena. However, due primarily to the variability of cloudiness and water vapor, the intensity of solar radiation at any particular time and place on the surface of the earth cannot be predicted. As solar radiation penetrates the atmosphere it is partly scattered or absorbed by clouds, molecules of air, water vapor, ozone, carbon dioxide, dust and pollutants. Both direct and diffuse radiation are ultimately functions of these variables.

The availability of solar energy can only be described statistically over long periods of time. The utilization of solar energy wil require the development of such long term data. But, as in any use of statistical information, a designer has to be aware of the extremes and the assumptions behind the data so that he uses his judgement in manipulating it to better fit his particular situation.

The amount of solar energy that one can expect can best be arrived at through long term statistical averages. These averages have to be developed over many years so that particular years of extreme high or low solar insolation are not used as the basis for designing a solar installation. The United States Weather Service has been keeping such solar insolation data at various stations (approximately 115) for varying numbers of years. The information is recorded as total insolation, direct and diffuse, falling on a horizontal surface. The average, total, daily radiation expected each month for each of these stations, and the number of years the data is based on are presented in Appendix A. Maps with gradient lines of insolation have been extrapolated from these data and are also in the appendix.

The insolation data are generated by a thermo-

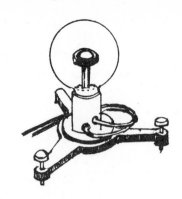

Figure 2.21

electric device called a pyranometer. The type of pyranometer used for recording total direct and diffuse radiation has a blackened, heat absorbing surface enclosed in a glass globe (Fig. 2.21). This device generates an electric current in proportion to the amount of heat it absorbs. If the device has been accurately calibrated, then the electrical output is compared to the output expected from the solar constant, or some intermediate standard, and recorded. Up until 1962, insolation values that were reported were based on the 1956 International Solar Constant of 1.94 gm cal/cm^2 min. Beginning in 1960 or 61, the recordings were based on a solar constant of 2.0 gm cal/cm^2 min.

In 1961 the Weather Service began a systematic replacement of older pyranometers which had used lamp black as the absorbing surface. However, the absorbing surface of the new pyranometers turned out to be thermally unstable, ultimately turning green in the last stages of degradation. The rate of degradation of any one of the devices depends on the climate and how much sunshine it receives. So the calibration of these devices—all across the U.S.—is off somewhere between (+)7% and (−)10%.[22] As a result, the Weather Service cautions against the use of their recent data; and, since 1962, have stopped publishing it. Slowly, as funds permit, they are replacing the defective pyranometers. The data, however, is available on magnetic tape or punched cards from:

National Climatic Center
Federal Building
Ashville, North Carolina 28801

One final, interesting note: Since the establishment of 2.0 gm cal/cm² min. constant, new investigations by NASA indicate that the value is 1.94 gm cal/cm² min ±1.5% (429 Btuh/ft²). This agrees with the former value for the solar constant before it was changed to 2.0 gm cal/cm² min. ±2%. Both the 1.94 and the 2.0 values are listed in the 54th Edition (1973-74) of the CRC **Handbook of Physics and Chemistry**.[23] Using the 1.94 values will yield a reliable average. However, if one wants reliable values at the level of 2.0 langleys/min., a simple ratio between 1.94 and 2.0 can be applied to the reliable 1.94 langley/min. data to produce reliable 2.0 langley/min. data. (The values are artificially constructed from electrical outputs, so the factors used to construct the data can be altered without altering the reliability of the data.)

NOTES

1. Statistics on the sun have been compiled from:

   R. C. Weast, Editor, **CRC Handbook of Chemistry and Physics 49th Edition** (Cleveland, Ohio: The Chemical Rubber Co., 1968-1969), pp. F144-145.

   J. L. Threlkeld, **Thermal Environmental Engineering** (Inglewood Cliff, New Jersey: Prentice-Hall, Inc., 1962), p. 280.

   B. J. Brinkworth, **Solar Energy for Man** (New York: John Wiley and Sons, 1972), pp. 25, 27.

2. B. Y. H. Liu and R. C. Jordan, "Availability of Solar Energy for Flat-Plate Solar Heat Collectors," **Low Temperature Engineering Applications of Solar Energy** (New York: ASHRAE, 1967), p. 1.

3. J. C. Threlkeld, **op. cit.**, p. 282.

4. G. I. Trewartha, **Introduction to Climate, 4th Edition** (New York: McGraw Hill Book Co.) p. 16. Figure adapted.

5. **Ibid.**, Figure from p. 13.

6. **Ibid.**, Figure adapted from p. 20.

7. **Ibid.**, Figure from p. 22.

8. **Ibid.**, Figure from p. 21.

9. **Ibid.**, Figure adapted from p. 16.

10. Brinkworth, **op. cit.**, pp. 28-29.

11. **Ibid**., p. 76.

    For a more complete treatment of atmospheric thermal dynamics, see "Heat Balance of the Atmsophere," **General Meteorology** (New York: McGraw Hill Book Co., 1959), pp. 25-44, by H. R. Byers.

12. F. Daniels, **Direct Use of the Sun's Energy** (New York: Ballantine Books, 1975), p. 17. Figure used.

13. Trewartha **op**. **cit**., Figure adapted from p. 16.

14. H. R. Byers, **op**. **cit**., p. 215.

15. Trewartha, **op**. **cit**., p. 12.

16. **Ibid**., p. 13.

17. **Ibid**., p. 22.

18. **ASHRAE Handbook and Product Applications: 1974 Applications** (New York: ASHRAE, 1974), pp. 59.4-59.5. Adapted from tables.

19. F. Daniels, **op**. **cit**., p. 33.

20. B. Y. H. Liu and R. C. Jordan, "The Long Term Average Performance of Flat-Plate Solar Energy Collectors," **Solar Energy** (Vol. 7, No. 2, 1963), p. 53.

21. Trewartha, **op**. **cit**., p. 14.

22. Telephone conversation with Mr. Himberger of the National Climatic Center, Asheville, North Carolina.

23. R. C. Weast, Editor, **CRC Handbook of Chemistry and Physics 54th Edition** (Cleveland, Ohio: CRC Press, 1973)

    Solar Constant = 135.30 mW cm^{-2}
    (p. F185) = 1.940 cal min^{-1} cm^{-2}

    Solar Constant = 2.00 cal min^{-1} cm^{-2}
    (p. F191) = ($\pm 2\%$)

CHAPTER 3
SOLAR ENERGY COLLECTORS

GENERAL COMMENTS

There have been many efforts to categorize the uses of solar energy. The use of solar energy has been broken down into the approaches to using it for space heating and cooling:

Passive and Active

It has been broken down by the end form of energy produced:

Solar Thermal and Solar Electrical

It has been broken down into manageable R&D programs investigating various intermediate processes as well:

Wind, Ocean Thermal Energy Conversion (OTEC), and Biomass

Hydropower is not generally discussed as a form of solar energy, but it is. It represents stored solar energy—in its potential to generate power. All of the above energy sources—and others—represent what have been broadly categorized as alternative energy sources.

The intent here is to deal with Space Heating and only tangentially with Space Cooling. Therefore, of the broad categories mentioned, Passive and Active Solar Space Heating have the most relevance to this discussion.

The earth is a solar collector, so are the other planets and asteroids in our star system. Everything under the sun is a solar collector. Approaching the concept of "Solar Energy" from this perspective has a liberating effect on the design process.

However, "a collector" has come to mean something quite specific. The public perception has been shaped so that it is generally thought of as "something on the roof." To some extent, this preformed image has made the concept of "Passive" solar heating somewhat inconsistent with the prevalent, preconceived notions of what a solar heating system should be. The following definition is instructive:

> A Passive Solar Heating or Cooling System is one in which the thermal energy flow is by natural means.[1]

To overcome any preconceptions, the concept of "a collector" will first be discussed at as general a level as possible. With this as background, it should be apparent that there is a continuum of possible collector options, each of which has its appropriate application.

Generally, there are two basic elements to any solar energy collector: the part that does the energy absorbing, and an ancillary piece, called a coverplate, that aids in capturing the energy available in sunlight. A car parked in the sun, is a good example of a collector. It becomes very hot in the car because the seats and interior walls absorb the sunlight and because the rolled up windows, which are very good coverplates, capture and hold in the heat that the interior absorbs.

The coverplate is the element responsible for creating what is often called the "greenhouse effect"—the trapping of heat (Figure 3.1). A good coverplate has several notable characteristics, or functions:

A. To minimize the loss of heat which has been captured.
B. To transmit as much solar energy as possible to the absorber.
C. To shield the absorber from direct exposure to the elements.

A. The coverplate, generally glass or plastic, reduces heat loss in several ways. It cuts reradiated losses by reflecting the longwave radiation (heat) back to the absorber, and it cuts convective losses by keeping the wind off the absorber plate. It should be noted that the nature of an absorber is such that it is not only a very good heat receiver, but that it is also a

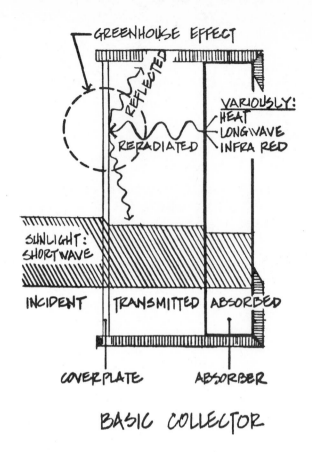

BASIC COLLECTOR

Figure 3.1

very good heat rejector. And, because it does reject (re-radiate) heat, a coverplate, which is opaque to longwave radiation (heat) is used in order to make the collector as effective as possible.

B. If there is a cover over the absorbing surface of a collector, it stands to reason that it must block some of the solar energy as the light passes through. Therefore, a good coverplate is one that lets in as much of this light as possible while satisfying requirement A, that it also holds in the heat coming back off of the absorber.

C. Sometimes, the performance requirements specified for a collector demand that exotic types of finishes be applied to the surface of the absorber to lower its tendency to reject heat. These finishes can be severely damaged just by the normal weathering processes; so, it is important that an absorber be protected by a cover. It would also follow from this, that the coverplate itself should be either very resistant to weathering forces or that it should be easily and cheaply replaced if it does not weather well. These are the basic criteria that a coverplate must meet; but economics will ultimately be the severest yardstick against which all the above (A, B, C) requirements will be measured. The main decisions required of a designer with respect to coverplates are: how many coverplates do I need (if any)?; and, what materials are best suited to my application? A discussion of these questions will come later.

The absorber is the heart of the collector. It absorbs the solar energy which hits it; and, in the absorbing process, alters the state of the energy to heat. This simply means that the absorber warms up when the sun shines on it. Like the coverplate, the absorber has several basic functions (Figure 3.2).

A. It should be very effective at absorbing incoming solar energy.
B. It should be very effective at conducting the heat away from itself.

A. Contrasted with the coverplate, which is supposed to have high transmissivity and low absorbtivity, the ideal absorber would have very high absorbtivity, very low emissivity, and very low reflectivity.

B. However, the absorber should have very good thermal conductivity. The ability of the absorber to conduct heat through itself and transfer it to other materials, or media, is very important in the utilization of collectors.

It is not likely that all of these characteristics are indigenous to any one material, but by a process of layering one can effectively build up an absorber that

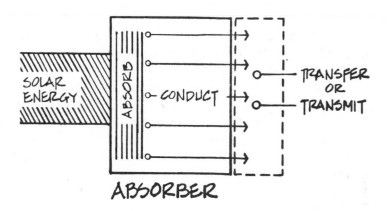

Figure 3.2

will have these characteristics. What particular mix of talents one builds into an absorber—as well as the whole collector—depends very much on the kind of application it is designed for.

The discussion so far has been at the broadest possible level in order to outline the basic elements of a collector. The automobile was mentioned earlier to show that many objects and their situations are good examples of solar collecting devices. The word **situation** is important because while the effective utilization of solar energy is not necessarily scientifically sophisticated, it is complex and should always be firmly referenced in its particular context.

41

This general discussion can now be used as background for viewing the various applications of both the Passive and Active Space Heating modes. Although these two modes represent distinctly different approaches to solar heating, there is a broad, blurred region where they overlap. The division into these categories, while instructive, is still quite artificial. With time the Hybrid Systems (that blurred area) will probably have more cases to its credit than either the Passive or Active Systems.

The Passive and Active modes can be distinguished from each other by the characteristics that in the Passive mode the collector often also serves as the storage and the distribution system. It collects the solar energy (Figure 3.3), stores the energy as heat and then it yields the energy back to its surroundings: the house itself is the solar heating system. The in-

tent in an Active system is to segregate the functions of collection, storage and distribution. Within physical and economic constraints, then, these separate functions and the equipment associated with each task are supposedly "optimized" to yield the best overall performance. The collector in such a system is designed to absorb the solar energy and to rise high enough in temperature to transfer this thermal energy to a working fluid (typically air or water) and then to storage or to the house heating system.

In terms of performance, Passive systems "coast." Typically, large building masses are involved for storage and the temperature rise required to store a great deal of energy is relatively small. The major limitation is the the range of temperatures (+8.3°C, 15°F) for storage is relatively small and has to be identical to the comfort range for space temperatures in order to directly produce "comfort" for the occupants. Active systems typically have relatively small amounts of mass for storage and, therefore, require large increases in temperature (+45°C, 80°) in order to store the same quantities of energy found in a similar Passive project. Because the energy in an Active system cascades down to space comfort conditions, the range for the large temperature rises has to begin substantially above the comfort range for space temperatures. Because the collector operating temperatures also have to cascade down to storage, this considerably elevated temperature range places the greatest limitations on Active systems: it is difficult to produce high temperatures in the winter and simultaneously collect large quantities of heat.

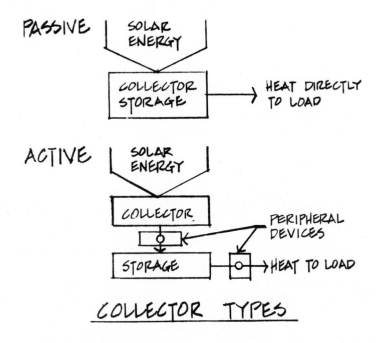

COLLECTOR TYPES

Figure 3.3

PASSIVE COLLECTORS

Figure 3.4 shows a section through a room in a typical home. Very simply, the home can become a solar collector. The room—its walls, ceiling and furnishings—is the absorber, soaking up and holding the warmth from the sun. The window is like a coverplate; if it is shut, it keeps the elements out and the heat in. In most homes, however, there has been no conscious design effort to treat the whole dwelling as a collector.

However, it has been known from antiquity that good southern orientations (in the northern hemisphere) take best advantage of the sun. Socrates (450 B.C.) was promulgating passive design principles—build with lofty south sides and squat northern exposures and use a generous southern overhang to shade the wall in the summer.[2] These principles were quite simply based on the observation that the sun is low in the sky in the winter and high in the sky in the summer.

In the late 1940's George F. Keck was designing passively heated solar homes in the Chicago area. The owners were using approximately half the fuel the utility company expected them to use.[3]

Large, south-facing windows are very effective during winter days. They transmit a lot of solar energy to the interior of the house. In fact, with a good day of sunshine, a room with large areas of southern glass exposure will probably over-heat. However, care must be taken to ensure that such large areas of glass do not lose so much heat through the night and on sunless days that they become an energy liability.

The sun will shine for about ten hours, yet heat will be lost through the glass for the full twenty-four hours of the day. More heat could be lost than gained if preventative measures are not taken. However, the designer can go far to utilize the natural solar and climate conditions to his advantage.

In fact, according to calculations by F. W. Hutchinson, unshaded south glass transmits 2.2 times as

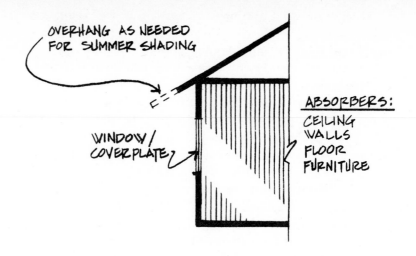

OVERHANG AS NEEDED FOR SUMMER SHADING

WINDOW / COVERPLATE

ABSORBERS:
CEILING
WALLS
FLOOR
FURNITURE

Figure 3.4

much solar energy in the winter as it does in the summer. If the glass is shaded to the extent that an overhang shades all of the glass at noon on June 21 and none of the glass at noon on December 21, then this factor rises to 2.9 times greater transmission in winter than in summer (seasonal averages).[4]

These ratios become even greater as the overhang is designed to exclude more summer sun and still not obstruct winter gain. The problem of excessive night heat loss can be eliminated by using insulating shutters or shades to substantially reduce the heat loss through the glass at night.

Still, the winter sun can overheat the space. Hutchinson (Purdue, Winter 1945-46) recorded temperatures of 26.6°C (80°F) with outside air temperatures of −1°C (30°F) and the house had been unheated. The sun had had to cycle space temperatures up from 1°C (34°F) over the 8 hour day. It was further pointed out that the house had very little capacitance or mass: the house could not usefully absorb the solar input. Although he pointed out that houses of different capacitance would produce different results. Hutchinson's conclusions[4] were in the form of a warning: that the direct use of

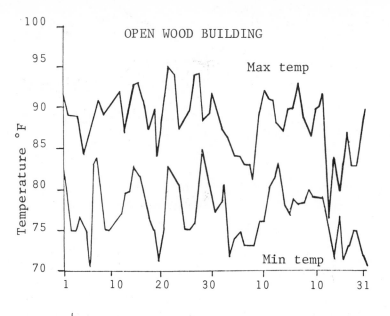

OPEN WOOD BUILDING

Max temp

Min temp

Figure 3.5

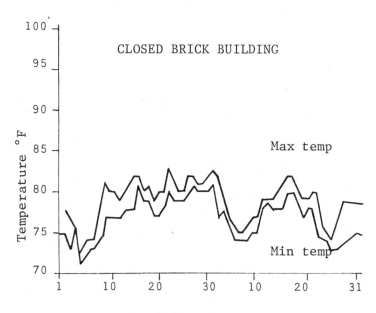

CLOSED BRICK BUILDING

Max temp

Min temp

Figure 3.6

solar energy in a house **in midwinter** may be undesirable, this condition would be acerbated in early and late winter.

Work done by Olgay and by the National Bureau of Standards suggests that the overheating would not be a problem if large amounts of building mass (capacitance) are introduced.

Figures 3.5 and 3.6 are from Olgay[5] and they clearly show the effect of capacitance. Because of high rates of heat transmission during the day the lightweight wood structure has to be opened to help dissipate the heat. This combination of factors results in the lightweight building experiencing high daily temperature swings inside the structure (13.9°C, 25°F). The capacitance of the mass and its accompanying low rate of thermal diffusivity has the effect of stabilizing the interior temperatures. The maximum daily excursion did not exceed 5°C (9°F) in the heavyweight building.

Figures 3.7 and 3.8 are also from Olgay.[6] Figure 3.7 shows the rates of heat transmission into the lightweight and heavyweight buildings. Over the entire day, they transmit approximately the same amount of heat, but heavy construction stabilizes the rate of flow with a process referred to as thermal lag. It is much easier to maintain comfort in the heavy construction example. In terms of mechanical assistance, the heavy construction would also be preferred because it has much smaller peak loads for air conditioning. This would result in smaller front-end equipment and lower utility rates—assuming there would be demand charges proportioned to the peak loads. Figure 3.8 is a record of the outside temperatures for the period of the experiment shown in Figure 3.7.

In other more controlled experiments, the National Bureau of Standards (NBS) has demonstrated the combined effect of massive construction with insulation.[7] In these experiments a small test house of solid concrete blocks was built in a warehouse-like test chamber. The temperature of the test chamber was

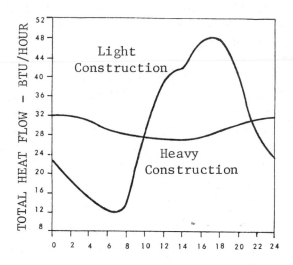

Figure 3.7

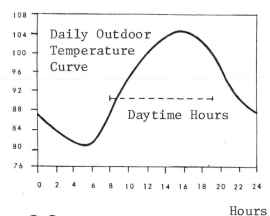

Figure 3.8

varied to simulate sol-air temperatures (Figure 3.9). In order to eliminate transient thermal effects, the test chamber was cycled through this temperature curve for three days; on the fourth day the experimenters recorded the temperatures inside the test house. This was the procedure for three different treatments of the test house:

1. No insulation.
2. 2" styrofoam on the inside surface of the blocks.
3. 2" styrofoam on the outside surface of the blocks.

Figure 3.9 shows the results of Case Three, insulation on the exterior of the mass. Compared to the outside temperatures, the inside temperatures were extremely stable. This is one of the results of having large amounts of mass inside the thermal boundary of the building. Another beneficial result is the phase shift of peak temperatures.

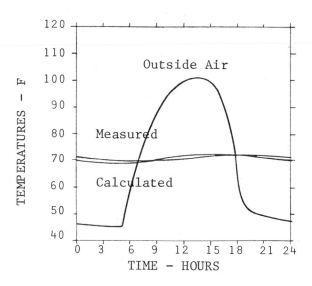

Figure 3.9

45

Figure 3.10 directly compares the three cases. In Case One, no insulation, temperatures varied approximately 10.5°F; in Case Two, two inches of styrofoam attached to the interior of the block, temperatures varied approximately 5.5°F; and in Case Three, two inches of styrofoam attached to the exterior of the block, temperatures varied approximately 2°F. It is instructive that the variation in Case One, 10.5°F, is quite close to the field measured variations reported by Olgay for a similar construction in Iraq during midsummer (9°).

It can be seen from Figure 3.10, that the greatest thermal comfort advantage is gained by insulating massively constructed buildings on the exterior. In terms of Passive Solar Design, the greatest gain also occurs from having large insulated quantities of mass (high capacitance) available to absorb the excess thermal energy that is often available—the mass functions as a thermal sponge. Ideally buildings, especially homes, should be built massively and they should be insulated on the exterior.

The preceding work provides some of the technical background for utilizing building design to help maintain space comfort and to conserve energy. Conservation results from the more stable temperatures, mechanical energy is not required to remove the unwanted dynamic impact of the environment—the building can do that.

A very good example of how buildings have evolved to conserve energy, based on climate and available resources is the vernacular adobe dwelling of the arid Southwest. It is well designed to provide comfort especially during the summer. It is also a good example of Passive Solar Design for cooling, based on the fact that the days are very hot and that the nights are quite cool. Trapping heat is not the goal, so there are no coverplates. Because it generally sits unshaded all day, the whole building is the solar absorber. Heat transmission to the inside is a real problem, so enough mass had to be designed into the surfaces of the building to provide a lag time that would keep the inside comfortable during the day for

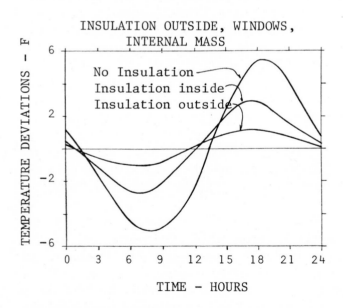

Figure 3.10

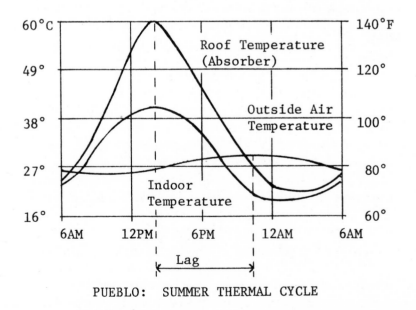

PUEBLO: SUMMER THERMAL CYCLE

Figure 3.11

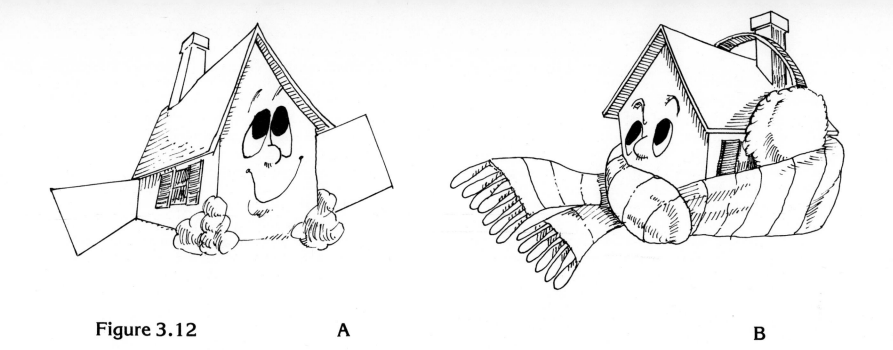

Figure 3.12   A   B

refuge from the heat. During the evening and night comfort is obtained out-of-doors in the cool night air. Similar building strategies and use patterns exist through out the Middle East and Mediterranean areas.

To illustrate the specificity of design evolution, the roof and the south and west walls are generally thicker than the east and north walls because they bear the full brunt of the sun during the day. They are designed to provide a lag of approximately 10 hours (Figure 3.11).[8,9] As the sun sets the walls begin radiating to the dwelling (absorber-in-reverse) until the peak daytime heat comes driving through around 10-11 p.m. It was common practice to sleep on the roof under a blanket and the stars.

The adobe dwelling is an example of a "passive architecture" dealing as best it can with a dynamic environment—passively. The adobe pueblo does quite well, however, dwellings of more recent vintage have survived to provide comfort only by their ability

to mechanically (actively) eliminate the dynamic impact of the environment. Energy is used to provide a static internal environment against the natural and dynamic forces of the outside environment. What is needed are building solutions that are not static: the building itself should be dynamic to better cope with the dynamic forces of the exterior environment. The building should open up on sunny winter days and close up on cold winter nights (Figures 3.12 A & B). A good passive design will give the occupant dynamic control of how the building deals with the environment to better provide comfort—naturally.

The following discussion of specific solar passive projects will proceed within the framework of the concept of dynamic passive controls. Each project discussed will provide examples of various levels of control. Each project is also noteworthy because it has become the archetype for others to generically follow.

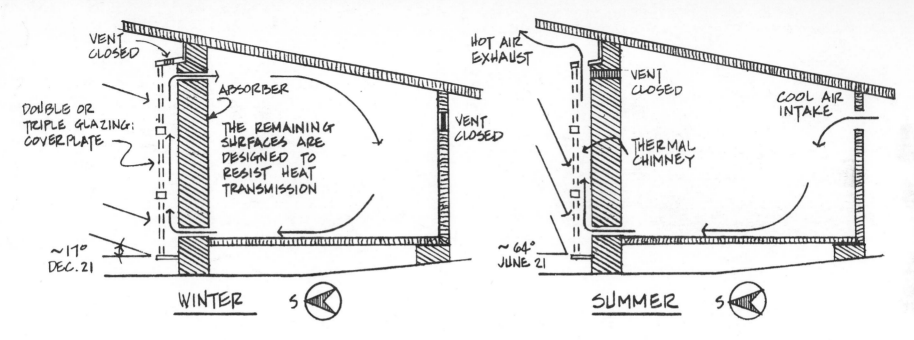

Figure 3.13

Trombe Wall.

Figure 3.13 shows a section through a passive collector-distribution system built by Jacques Michel in Montmedy, France in 1970.

This design is referred to as a Trombe Wall after Dr. Felix Trombe of the French CNRS (National Center for Scientific Research) who is responsible for the concept.

The concept is that most of the house is of lightweight highly insulating materials, and that the mass normally associated with the passive house is concentrated in a single south-facing glazed wall. This wall is the regulating, or controlling, mechanism. As it absorbs the solar energy, the exterior surface quickly rises in temperature. This relatively high temperature drives two mechanisms: one, it sets up a thermosyphon convective air loop with the cooler house since the exterior surface can reach temperatures in excess of 60°C (140°F);[10] and, two,

it charges the mass wall with heat, slowly driving the solar input through the mass wall. The wall has a thermal lag such that the wall becomes a radiant heating panel only at night after the sun has gone down and the exterior temperature of the wall has dropped too low to thermosyphon heat into the house.

Figure 3.13 A illustrates the winter operations. During the sunlit hours heating the house is accomplished by natural, thermal convection. Cooler air enters through a vent at the base of the 45 centimeter (17.73 inches) black, concrete wall, passes between the concrete and the glass and flows as heated air through a vent at the top of the wall. For a single-story house of approximately 100 square meters (1076 sq. ft.), the air cycles completely in an hour. At night the vents are closed to prevent a reversal in the direction of the thermosyphon loop, which would heat the great out-of-doors, and heating proceeds by radiant heat transfer.

Montmedy is between 49 and 50° North latitude—about the same as Winnipeg, Canada. Michel estimates that a house this size (280 cubic meters of living space) requires 7000 kilowatt hours for space heating annually. At Montmedy, 5400 kilowatt hours were supplied by solar heating and the remainder through an auxiliary electrical system. The annual heating cost for electricity was approximately $225 as compared to an estimated $750 for a home entirely heated by electricity in that area. This amounts to a 77% reduction in heating load and a 70% reduction in the cost for a winter's heat.[11] Rough design guides suggest that there should be one square meter of Trombe Wall for each 10 cubic meters of house volume.

Ventilation and some cooling effect in summer is achieved by shutting the return vent to the house and exhausting the heat in the collector, through convection, to the outside. The collector functions as a "chimney" (Figure 3.13 B) and the open vent allows the collector to draw air from the house, enhancing the ventilation aspects of the dwelling. The main thrust of this design, however, is to provide heat for the winter. Whether the summer ventilation technique would have general application for the amelioration of uncomfortable conditions in regions closer to the equator is open to question. As will be seen in the following projects, it can be very important to insulate against summer heat gain. However, the Michel project demonstrates a high degree of effectiveness with little or no technological support, showing the design can directly impact the energy requirements of a dwelling.

SkyTherm.

Another generic type of Passive Solar Heating and Cooling is the SkyTherm House concept by Harold Hay. As in the Trombe case, mass is interposed between the house and the sun. Harold Hay has built a house in Atascadero, California which incorporates a system that can be used to supply and remove heat

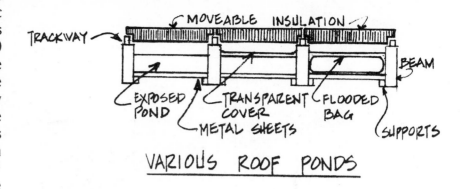

VARIOUS ROOF PONDS

Figure 3.14

from the building with a simple reversal of basic processes. The collector is a "solar pond" (Figure 3.14) on the roof of the house which can be covered with insulating panels. This roof supports shallow plastic bags inflated with water. With six inches of water one has the heat storage capacity of one foot of concrete, but one has it for one-fourth of the weight— a very important structural consideration for a roof.

During the winter, when heating is the objective, the insulating panels are removed during the day so that the water can absorb the solar energy. At night the insulating panels are replaced to hold the heat. Then, throughout the night, the heat is conducted through the metal ceiling and radiated to the space (Figure 3.15 A).

During the summer, the panels cover the water pillows during the day. This shields them from the outside temperature and the sun's radiation, but heat flows from the dwelling through the metal roof and it is absorbed by the water. The insulating panels are removed during summer nights, and the water radiates its heat to the clear night sky. If the bags are flooded, the evaporation of water works to enhance the cooling effect. This process cools the bags of water so that they can soak up heat from the dwelling during the following day (Figure 3.15 B).

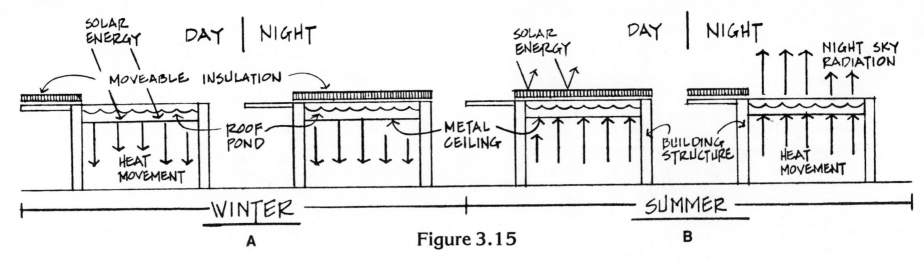

SOLAR ENERGY · DAY | NIGHT
MOVEABLE INSULATION
ROOF POND
METAL CEILING
HEAT MOVEMENT

SOLAR ENERGY · DAY | NIGHT
NIGHT SKY RADIATION
BUILDING STRUCTURE
HEAT MOVEMENT

WINTER

A

Figure 3.15

SUMMER

B

Hay's design provides a cooler roof than normal in the summer; however the cooling effect comes from the fact that the human body will radiate its heat to the cooler ceiling. During the night, as one sleeps, the cooling effect can be controlled by the insulation. If the insulation is removed, the ceiling will become cooler. If the insulation is left in place, the ceiling is kept warmer. It is claimed that the temperature of the ceiling can be controlled by selectively removing specific insulation panels at night. In this way different comfort levels can be created in one room. If half the ceiling in the bedroom is uncovered, people on opposite sides of the bed will experience different cooling effects; the person under the covered side will feel warmer than the person on the uncovered side, since the one under the insulation will radiate less heat to the warmer surface which is closer to them.

This design provides 100% of the heating and cooling needs of this house—in Atascadero, Calif. Caution, however, must be exercised in adapting these principles to other, more severe climates. Temperatures only infrequently go below freezing in Atascadero, so the water design is quite appropriate and takes maximum advantage of the nuances of its microclimate.[12]

The element of control involved in making the building dynamic is the roof pond insulation. The occupant, through experience, learns just how many hours to leave the insulation open on a bright sunny day: long enough to collect enough solar energy to carry through the night; but not so long that the system would overheat the house.

Drum Wall.

Steve Baer, who heads up a corporation called "Zomeworks," is another front runner in the application of passive utilization of solar energy. Baer's response is what he calls a "drum wall" (Figure 3.16). Discarded steel containers similar to oil drums (slightly smaller) are painted black on one end and are filled almost full with water (this allows some room for the thermal expansion of the water). The drums are then stacked horizontally behind two panes of glass on a south facing wall with a blackened bottom facing the outside.

The Baer design is quite similar to the Trombe/Michel example. The design is a highly insulated, lightweight building system to which a concentration of mass is added—as a whole building module—to

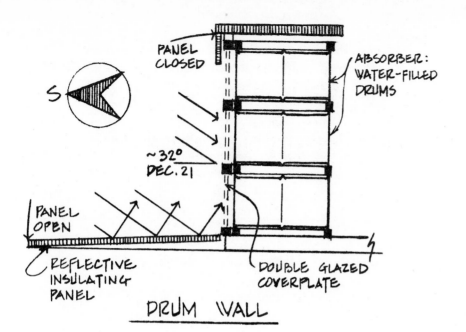

PANEL CLOSED

ABSORBER: WATER-FILLED DRUMS

S

~32° DEC. 21

PANEL OPEN

REFLECTIVE INSULATING PANEL

DOUBLE GLAZED COVERPLATE

DRUM WALL

Figure 3.16

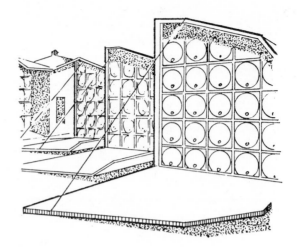

Figure 3.17

the south side of the dwellings (Figure 3.17). Like Michel, Baer depends on a combination of convective air movement and radiation to circulate warmth throughout the dwelling; and, in the event that there has been insufficient collection of solar radiation, he has a fireplace which provides supplementary heat.

Baer has a moveable insulation system that the Michel-Trombe wall does not. Outside the glass is a sandwiched insulating panel of aluminum skins with a three inch thick core of styrofoam. The panels are nine feet square and are designed to be lowered during the winter days to expose the drums to solar radiation. These panels work to increase the solar effect since their highly reflective surfaces help to increase the amount of energy actually reaching the drums. At night the panels are drawn up to insulate against heat loss to the outside. With the panels raised, the drums transfer most their heat to the interior.[13]

The significance of the Baer design lies in the elegance of utilizing the dynamic insulating system to actually amplify the amount of solar energy available for space heating. This is an added nuance compared to the relatively straightforward insulation of the solar roof pond. In addition, by demonstrating that a mass module can be virtually just clipped onto the south side of a dwelling to achieve passive solar space heating, this example is a prototype for the passive remodeling of the large number of lightweight frame homes that have relatively good southern exposures. With 60 million existing housing units the Baer design may well have a significant role to play in converting existing homes to solar passive space heating.

Wright House.

The David Wright House (Figure 3.18) parallels the Hay and Baer examples. It is prototypical for what

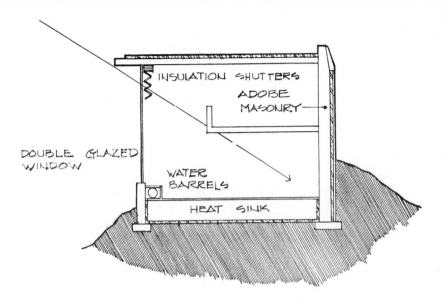

INSULATION SHUTTERS

ADOBE MASONRY

DOUBLE GLAZED WINDOW

WATER BARRELS

HEAT SINK

Figure 3.18

is called a "Direct Gain" Passive Solar Space Heating design. The direct gain refers to the fact that the solar energy is accepted directly into the space through windows. In this regard "Direct Gain" is simply a revival of the Keck homes of the early 1940's.[3] However, direct gain solar systems generally imply that other efforts are also undertaken to optimize the use of the solar gain.

In the direct gain design there is no mass wall that intercepts and stores the sun's energy before it enters the room. Typically the floor is designed to be a large portion of the mass required for storage, since it receives the full brunt of the solar radiation; however, massive structural walls are also used as part, or all, of the passive thermal storage system. In the Wright House, the adobe floor is approximately two feet thick and it is underlayed with insulation; in addition, the North, East and West walls are made of adobe that is insulated on the exterior. The massive walls

and floor absorb the incoming solar radiation and stabilize temperatures in the thermal comfort zone of people. At night, insulating shutters are lowered to reduce the heat loss back through the glass. The Wright House gets approximately 90% of its heating requirements from solar energy in a climate that has an average of 6000+ Degree Days per winter.[14] [15]

Unlike the previous mass wall/roof systems that are first heated by the sun before the energy is released in the space, the direct gain can be comfortable at much lower temperatures—the occupants can bask in the direct radiant energy from the sun without space temperatures being high enough to provide comfort by themselves. Direct gain systems, because they do not develop relatively high temperatures at the skin of the building like mass walls do, have correspondingly lower heat loss rates back to the cold outside environment. Therefore, they have relatively higher overall system efficiencies: proportionally more solar energy goes to meeting demands for heat.

Compared to Indirect Solar Passive Heating examples like the Trombe or Drum Wall, there is less specific control—by design—over space temperatures in a Direct Gain approach. Because Indirect Gain is "linked" to the house, it can be "valved" off by closing dampers to shut off convection or by using heavy curtains to reduce radiant transfer. Overheating occurs outside the comfort zone of the house, but still within the boundaries of the system: heat is retained. A Direct Gain system copes with overheating by venting the excess heat in order to maintain comfort. In a system designed with sufficient internal mass, however, this venting will be a rare occurrence, limited to the early and late winter periods when the heat would not be needed to meet demand requirements; and it should not affect overall seasonal system performance.

The Davis House.

This house represents a slightly different wrinkle

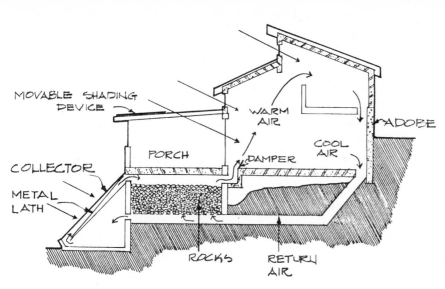

MOVABLE SHADING DEVICE

WARM AIR

ADOBE

COLLECTOR

PORCH

DAMPER

COOL AIR

METAL LATH

ROCKS

RETURN AIR

Figure 3.19

on passive design. All energy flows, in this design by Steve Baer, are by natural means (Figure 3.19). However, the collection and storage of solar energy are outside the "boundaries" of the house. This design approximates an active solar heating system, but Nature handles the movement of energy without mechanical assists from fans or pumps. The amount of heating accomplished is occupant regulated by adjusting the dampers to throttle or re-direct the flow. Paul Davis has suggested three modifications to improve the design: one, move the rock bin under the house to reduce external heat losses: two, integrate a thermal chimney with the collector to induce thermosiphon convective cooling in the summer; and, three, add a greenhouse to boost natural humidification during the winter.[16] This system has been credited with providing 75% of the space heating requirements for the 1000 square foot home.

Another approach to Passive Space Heating that merits specific mention is the Solar Greenhouse. The concept is that a greenhouse is quite good at collecting solar energy; so, why not simply attach one to the south side of an existing or new home so that excess heat generated during the day can be tapped for use in the house. At night, the greenhouse is thermally isolated so that it does not add to the heating load. The basics of operation revolve around taking advantage of the solar energy that may be available—if it is needed. Elaborate schemes for storing thermal energy are counter-productive in these elegantly simple, low-cost solar collectors. For a more complete development of solar greenhouse principles see the work done by Bill Yanda and Rick Fisher or by James C. McCullagh.[17]

In the preceding examples there resides most of the basic principles that are utilized in Passive Solar Space Heating and Cooling. These designs have been designated "prototypical" by the fact that they were among the earliest examples in the recent development of solar energy for space heating. Since they were built there have been hundreds of homes built that are variations on one or more of the above designs. Passive design principles, it should be noted, are also finding their way into commercial and industrial buildings: they are not principles that are limited to residential use.

Good Passive Solar Design is beneficial and cost-effective in all climates. The principles can be designed in from the beginning for very little in terms of added cost. Work that Londe-Parker-Michels, Inc. has been involved in suggests that it is economically reasonable and technically feasible—using passive design principles—to build homes that require only 30% of the energy for heating that is used by the average home built in 1976.[18]

Passive Solar Design for existing buildings, while not as cost effective as passive solar in new construction, will still yield a reasonable economic payback. A

a. Insulated brick wall with stucco finish.
b. Existing brick or stucco.
c. Stucco & glass bay to be changed to glass & wood frame.
d. Glazing mounted 10" from brick wall. 2nd floor only.
e. Solar greenhouse. Erected each fall, removed each spring.
f. Wood framing structural units for all glazing materials.
g. Storm windows and storm door for front porch.
   Winter use only.

WINTER CONFIGURATION

Figure 3.20

Retrofit Passive design executed by Londe-Parker-Michels, Inc. (1978) took a forty-year old masonry duplex (Figure 3.20) and insulated the masonry walls on the outside and constructed a Sunwall (275 ft.²) that consisted of a winter-only greenhouse and a 2nd floor glass wall. The conservation efforts are expected to reduce demand 60% and the Sunwall is expected to reduce demand an additional 20%. The anticipated payback period is 10 years.[19]

The LPM experimental retrofit is fully instrumented and should be a continuing source of design information in the near future.

Besides being cost effective—yesterday—to employ solar passive space heating in new construction, Passive Design is a "new" technology that can be embarked on immediately: there are no commercial barriers. Because Passive efforts merely represent "good" design, there are no real code problems or jurisdictional disputes among the trades. The technologies involved are already part of the building inspector's repertoire. Banks are also quite willing to finance "good" design—because there will be at least "three comparables" that an appraiser can refer to. In short, there are no barriers to Passive Solar Design. Ironically, the only threat to Passive Solar would be the possible implementation of a prescriptive type energy conservation code that would restrict south wall glass.

Passive Solar Design is also relatively liability free. The designer really does not bear any additional liability over and above what he would be assuming for the building anyway—the building **is** the solar system. The design consequences need not be too forbidding either. Although one would like to be exactly right all the time, Passive Design is quite "tolerant" of oversights.

In conclusion, Passive Solar Design will become a way of life for the design professions. It offers greater potential and challenge for creativity; its aesthetics are generally quite pleasing; its low initial cost will yield good paybacks; it is relatively barrier-free in the marketplace; and it will need to be done even for active-type systems to help make them more cost effective. As passive approaches develop they may challenge the need for active type space heating systems at all—especially for new construction.

ACTIVE COLLECTORS

There are two basic types of active collectors. They are commonly referred to as "flatplate" and "concentrating" collectors. Each type has its own particular set of appropriate applications.

The flatplate collector has widespread applicability. It can be used for preheat in industrial processes, for residential and commercial heating and air conditioning and for domestic hot water. Because it is appropriate to such a large sector of the built environment, the flatplate collector (FPC) will be discussed at length here. Since everyone is familiar with a residential setting, space heat for residences will be the primary vehicle for illustrating the principles involved in using FPCs.

Since there are a variety of FPC systems already developed and available, off-the-shelf systems are cropping up in Sweet's Catalog. In the near future architects will be heavily involved in all manner of solar-type systems. The architect, with his responsibility for the overall success of a project will have to take the lead to ensure that the solar system is appropriate: the architect cannot rely on the engineer alone to make this assessment. The engineer is not responsible for assessing the impact of the different solar design options on the architectural design, and the solar system definitely should not be designed-in **after** the building design has been fixed.

If the architect intends to successfully utilize solar energy, in an active sense, he has to be more than passingly familiar with the options and the decisions that are implicit in the options. Therefore, the thrust of the discussion will be to cover the options and their implicit decisions and to cover the impact of the different systems on architectural design. The designer will then be able to select the system best suited to his needs. This discussion will also be of value to the design engineer, since the information available to professionals generally comes from the trades and does not always reveal the implications involved in selecting "their equipment."

Concentrating collectors find their greatest applicability in research settings and in large scale power production schemes. They have little applicability for residential use because they are designed to produce very high temperatures. For these reasons the principles of concentrating collectors and their implications for architecture are discussed in Appendix C.

With the active collector, collection and storage of heat become separate operations which are linked by a heat transfer media (Figure 3.21). The purpose of this media is to remove heat as effectively as possible

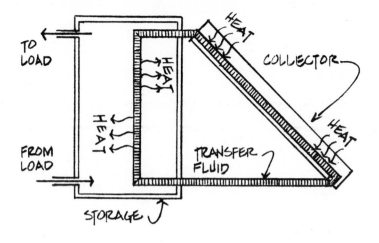

Figure 3.21

from the collector and to transport it to a storage device where it is held for use at a later time. This transfer process sets some very real constraints on the collector design. Besides a coverplate and an absorber, an active collector generally has tubes or fins channeling the heat transfer media from an inlet to an outlet; insulation which minimizes heat loss from the back of the collector (the side not exposed to solar radiation); and a container, or casing, which holds the coverplate(s), absorber, tubes or fins and insulation together (Figure 3.22).[20]

The purpose of the tubes or fins, besides their directing the fluid, is to provide sufficient surface area for heat transfer so that the heat trapped in the absorber can be effeciently transferred to the fluid which will take the heat it picks up to the storage unit. The amount of heat absorbed into the fluid is dependent on its specific heat, cal/(gram·°C) (Btu/lb·°F), the rate of flow of the fluid (kg/hr) or (lb/hr), the temperature of the absorber, and the temperature of the fluid when it enters the collector. (See the appendix: "Concepts of Heat Transfer"). These factors will

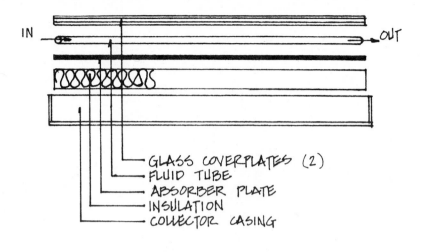

GLASS COVERPLATES (2)
FLUID TUBE
ABSORBER PLATE
INSULATION
COLLECTOR CASING

Figure 3.22

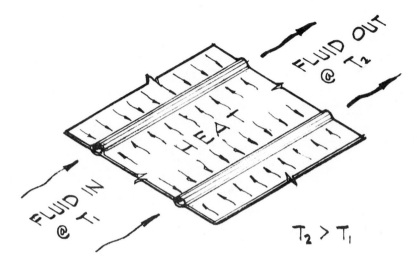

FLUID OUT @ T₂

FLUID IN @ T₁

$T_2 > T_1$

Figure 3.23

determine the outlet temperature of the heat transfer fluid; and it is the difference between the inlet and outlet temperature that determines how much heat one has been able to collect.

Since active collection implies that storage is no longer among the functions of the absorber, high heat capacity is no longer an important characteristic of the absorber. However, it becomes very important for the absorber to be able to quickly conduct heat from one point to another (Figure 3.23). This represents a major shift in absorber function between the active and passive mode.

The basic function of the coverplate remains the same in active collection as in passive collection. It still must reduce heat losses to the atmosphere, protect the surface and transmit as much solar radiation to the absorber as possible.

The insulation, by reducing losses through the back of the collector and by reflecting (if it has a foil facing) heat back into the absorber plate, helps maintain high collector temperatures. It might also be of

benefit, depending on needs and economics, to insulate the sides of the collector, or to at least isolate them from the absorber plate so that the whole casing perimeter does not become a path for heat loss by conduction.

The Casing: The casing, besides holding all the other elements in the collector together, should also seal out dust and moisture. If significant amounts of dust should filter into the collector, and settle over the absorber it could seriously affect the absorbitivity. If dust should cover an inside coverplate, it could drastically reduce the transmissivity. Moisture infiltration should be kept to a minimum since a collector would act much like a solar still and create a layer of condensation on the inside of the coverplate. This condensation could seriously reduce the transmission of energy to the absorber.

Flatplate collectors usually are set up in one of two

Figure 3.24 A

Figure 3.24 B

places—either on the roof or the building or free-standing somewhere on the site. Occasionally they are integrated with or attached to a south wall. If the collector is freestanding, or if the roof is not at the angle one determines is best for collection, then the casing must have its own structural support system. For the sake of economy, the casing is usually the main structural element. However, the collector can be integrated with the structure of the roof, or of the wall, under certain conditions.

If a collector is mounted on a roof—especially if it is a retrofit operation where a building is being converted to solar usage—the weight of the collector with a full fluid load has to be carefully considered to ensure that the load carrying capacity of the roof is not exceeded.

If the residence is being built with a solar collection system designed into it from the start, there are additional options. The roof angle can be designed to match the optimum collector tilt angle. If this is the case, then the collector can be placed on the roof, in the roof or underneath the roof. Or, better still, the

collector can be the roof (Figures 3.24 A,B,C,D).

Figure 3.24 A shows a collector mounted on the surface of a roof. This has some advantages in flexibility, but there is some redundancy in placing it over a roof. If the collector area needed is smaller than the roof area, then it would probably be more advantageous to place the collectors on the roof rather than have to work out the details for two different roof sections and their interface. It is very likely that a regular roof with the collectors on top of it would actually be cheaper in this case.

Since collectors represent the largest portion of the cost in a solar collection system, the collectors might have to be acquired in several installments, expanding the system as funds are available. Choosing to mount on the roof, rather than in the roof, would certainly simplify the process of making additions. Since there is a large stock of existing housing, it seems that retrofit installations could be a much larger market than new solar building starts. This would make surface mounted collector systems the most widely applicable design. Therefore, manufacturers

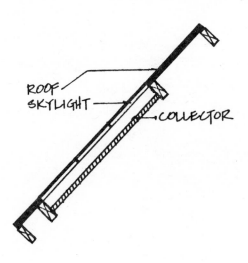

ROOF
SKYLIGHT
COLLECTOR

Figure 3.24 C

COLLECTOR/ROOF

Figure 3.24 D

will probably tool for this particular design first, making the surface mounted design the most available and, probably, the least expensive.

Figure 3.24 B shows a collector nestled between the roof rafters. The coverplates and the cap strips between each row of collectors forms the finished surface of the roof. In this design, the cost of the roof that one has eliminated can be credited to the cost of the collector. However, if the design dictates that the collectors cover a large area, special provision has to be made for reaching each module for servicing or replacement, since it is not likely that the coverplates would make a very desirable work surface.

Figure 3.24 C shows a collector that has been split. The coverplate serves as the roof surface similar to example 3.24 B, but the absorber and casing hang beneath the rafters. Basically, one designs a skylight —for which the technology is already well developed —and then one hangs an absorber below it. This approach eliminates a lot of detailing problems of putting the collector in the roof. This allows for plenty of

tolerance in manufacturing which would serve to reduce the final cost of the collector. It also takes advantage of the well developed industry for skylights; and it could answer the objections of those who reject solar energy on the basis of aesthetics, since it hides the collectors from sight.

Figure 3.24 D shows a particular collector that was developed at Los Alamos Research Center.[21] It was designed to replace the roof entirely. The collector is formed from mild steel in such a way that it is a structural steel channel capable of carrying its own and all other roof loads. This solution has a great deal of elegance because it has eliminated a lot of roof detailing problems. The manufacturing of these collectors is already automated and they can be installed very quickly. The models that have been designed so far range from ten to twenty feet in length. However, it is assumed that these units can be made longer if accommodating adjustments in the structural section of the collector are made.

There is some question with respect to all of these collector systems about possible problems with union jurisdictional conflicts during installation, especially with respect to the D option. One cannot really foresee what shape such conflicts might take or whether they will even exist, but one should be alerted to this possibility. It should also be noted in options B, C, and D, since they in some way form the roof, that any heat lost through the back of the collector actually is lost into the building, whereas in option A this heat is dissipated to the outside environment. This heat would be a boost to the heating system during the winter, but it would also be an added cooling load during the summer. The various alternatives have to be carefully chosen in the context of the climate of the project.

There are some major detailing problems that one should be aware of in collector design. There is the

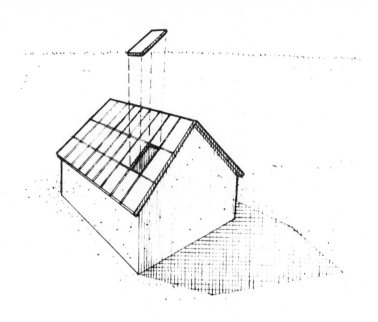

Figure 3.26

problem of modularity: how big should one collector unit be? The most frequent solution has been to keep the width of the collector to between two and three feet. This dimension is based primarily on using glass as a coverplate, which can structurally span these distances without intermediate structural supports.[22] With plastic coverplates, widths go up to four feet, but this seems to be the upper limit—probably due to tolerances required by thermal expansion and to the fact that various building industries already standardize around a four foot module. The length of collector modules is variable with manufacturers offering options up to twenty feet as standard. Ideally, one would prefer to have the length be the height of the collector array, which is typically determined by the designed temperature rise required between inlet

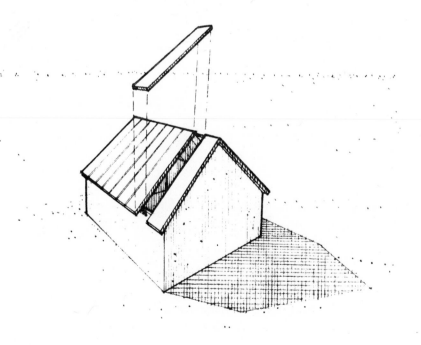

Figure 3.25

and outlet temperatures. If the module length is the height of the array, then one avoids the problem of connecting two collector modules in tandem. Each connection made requires additional hardware and multiplies the chances that the collector might leak. If the collector **is** the roof system, the roof would leak if the seals between collector modules deteriorate. The principal cause of leaks which occur at connections and seals is thermal expansion. The major elements of a collector become very hot, and, because the ideal materials have high conductivity, they invariably have high coefficients of expansion as well. The movement of parts as they expand and contract—often at different rates—is a serious problem.

Solutions to this problem seem to reinforce the trend of keeping at least one side of the collector unit short, since the tolerances that have to be absorbed as it expands and contracts are relatively small over short distances. The shorter collector widths seem to work well with respect to this thermal expansion and movement, but the movement of different parts of the collector over the long dimensions remains a problem.

With respect to expansion, collector configurations A and C work well since expansion can occur without interrupting the integrity of the roof. In A, the roof is underneath and apart from the collectors; the only problems it has are those of a normal roof. In C, the skylight-type roof should be as reliable as a normal skylight installation. However in configuration B and D, the problem of thermal expansion must be met head on and solved if using the collector as a roof is to be a viable solution.

It would probably be best to try to carry the entire distance of the collector array (Figure 3.25) with one collector unit and to absorb large tolerances only once at each end. If shorter collector units (Figure 3.26) are used, then a whole series of joint related problems are introduced. There have to be structural connections between collectors; multiple connections for the fluid have to be made between collector units; and all the joints between the units have to absorb expansion and still remain leakproof (Figure 3.27).

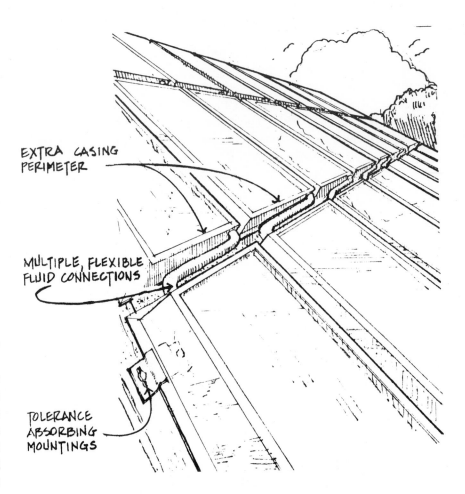

EXTRA CASING PERIMETER

MULTIPLE, FLEXIBLE FLUID CONNECTIONS

TOLERANCE ABSORBING MOUNTINGS

TOLERANCE MOUNTINGS REQUIRED WITH EACH MODULE.

CONNECTIONS BETWEEN MODULES MUST BE TOLERANCE ABSORBING AND LEAKPROOF.

Figure 3.27

Insulation: The most important consideration for insulation is its resistance to deterioration at relatively high temperatures. Some flatplate collectors are designed to operate as high as 176°C (350°F) for extended periods of time. However, one should realize that the stagnation temperature for most flatplate collectors is also in this range—about 150°C (300+°F). If the collector coolant (heat transfer medium) should ever fail to flow—especially during the summer — then the temperature of the collector could easily reach and hold these temperatures. It is not suggested that a collector be designed to operate at this temperature; but it should be realized that no matter what the design temperature is, if the collector coolant should stop flowing, the different elements of the collector will have to be able to survive these temperatures. It would be very costly in terms of time, energy, and money to have to replace all the insulation behind a collector if, after a short period of time at high temperatures the insulation fails. Styrofoam for example decomposes around 120°C (250°F). On the other hand, mineral fibers require protection from moisture so that they do not get soggy and pack down, loosing their cellular air pockets and their resistance to heat loss. Even if some insulations can withstand the temperatures, care has to be taken to make sure that they do not outgas—a process where volatile agents vaporize, migrate out of the insulation and condense on the inside of the coverplates. One has to specify high temperature grades of common insulations, like fiberglass or urethane, to ensure that this and other decompositon problems are avoided. Make sure the insulations are suitable for high-temperature usage.

Freestanding and surface mounted collectors require that the case be well sealed against moisture to prevent the degradation of the insulation due to water. With these installations, it is assumed that the insulation in the collector works only with the collector; the building requires its own separate insulation. Conservation of materials and economics come together at this point to at least reinforce some of the arguments for integrating the collector with the roof. If the collector is integrated with the roof, the insulation in the back of the collector can act as insulation for the building as well. With the elimination of redundancy in the use of insulation, some marginal cost savings may be effected. Considering such options will depend on whether or not the different costs of the various collector designs are close enough to attack insulation usage as a source of savings. For example, if surface mounted and integrated collector systems are estimated to cost the same, both initially and over time, then it might represent a saving to go with the integrated design and save the cost of the added roof and its insulation. However, if the extra detailing for the integrated design causes it to cost significantly more than the surface mounted design, then one should not expect to reap savings just by eliminating the extra insulation. The old saying "penny wise and pound foolish" would be operant here. The collector design as a whole should be considered in efforts to reduce costs, not just the insulation.

In the case of example D where the steel collector is used structurally, fireproofing might be required in some instances. If this were the case, it would be important to find a fireproofing with a good insulating value in order to eliminate the added cost of insulation.

Coverplates: The most widely used coverplate material is glass. However, various kinds of plastic sheets have also been used.

One of the biggest problems is the possible breakage of glass coverplates. In fact, the introduction of plastics as coverplates at all is due primarily to the search for a breakage resistant material for a coverplate. It should be noted that except for its breakability, glass is an ideal material. It is very effective at satisfying all other criteria set for a coverplate. In addition, glass is readily and economically available.

If breakage is expected to be a particularly severe problem then plastic coverplates, or combinations of plastic and glass, would be needed to solve this particular problem.[23] However, the plastics have some inherent physical limitations which restrict their utilization as coverplates. Foremost among these is the fact that the temperatures that a collector can reach—176°C (350°F)[24]—will destroy most plastic coverplates (Figure 3.28). Regular glass, in comparison, has a maximum operating temperature of 204°C (400°F).

So, depending on the kind of plastic chosen, one has to make sure in the design of the collector, that the stagnation temperature (temperature of the collector when the coolant is not circulating) does not exceed a temperature that the plastic sheets can handle continuously over the projected life of the collector. Alternatively, one could select a relatively inexpensive, thin plastic film (two to four mil. thick); accept the fact that it will last only a short time; and schedule routine replacement of the film-coverplate.

Aside from the temperature of the collector possibly destroying the plastic sheets or films, the plastics also have a tendency to degrade chemically under continued exposure to the ultraviolet portion of the solar spectrum.[26] The plastic will turn color, critically reducing the amount of solar energy getting through the collector. There are chemical treatments which can prolong or even prevent this ultraviolet degradation process, but they, necessarily, increase the cost of the film or sheet.

Another temperature related problem in using plastic materials is thermal expansion. The collector has to be designed to cope with the casing, absorber and coverplate all expanding at different rates (Figure 3.29). Aluminum, steel and wood are probable casing materials; aluminum, steel and copper are common absorbers; and glass and plastics are coverplates. The more disparate the expansion rates of the components, the more the tolerances involved will be a problem. This means a collector design which uses plastic coverplates would have to have details to cope with expansion that would be 8 times as large as a collector which utilizes glass. The relative numbers

	Thickness	Solar Energy Transmittance	Max. Operating Temperature	Infra-red Transmittance
Float Glass	3.18 mm *	.85	204 C	0.
Tempered Glass	3.18 mm	?	290 C	0.
Fluorinated ethylene propylene	0.05 mm	.97	210 C	.73
Methyl methacrylate	3.18 mm	.89	82 – 93 C	?
Polycarbonate	3.18 mm	.81	121 – 132 C	low
Polyethylene	0.05 mm	.92	65 C	.86
Polyethylene terephalate	0.05 mm	.84	105 C	?
Polyimide	3.18 mm	?	210 C	?
Polyvinyl fluoride	0.10 mm	.92	107 C	.30

* 1/8 inch

[25]Figure 3.28

MATERIAL	EXPANSION*
Aluminum	24×10^{-6}
Copper	17
Glass:	
Pyrex	3
Float	8.7
Methyl Methacrylate	73.8
Oak:	
Along fiber	5
Across fiber	54
Polycarbonate	67.5
Steel	12

* inches per inch per °C

[27]Figure 3.29

for glass, steel and plastic would also show that if the collector casing and absorber were made from steel, then there is a better material match between the expansion of steel and glass than between a steel and plastic combination. However, using a similar set of comparisons with respect to breakage, if glass is 1.1, then plastic is about 400 and steel is 2000.[28] These numbers gauge their relative resistance to fracture.

Some consideration has to also be given to the structural load carrying capability of the different coverplate materials. For flat sheets under moderate wind load and dead load, plastic manufacturers have published nominal maximum sizes recommended for the short dimension of plastic sheets. For "plexiglas"[29] (polymethyl methacrylate) and for "lexan" (polycarbonate) the recommended short dimension of a 3.2 mm (⅛") sheet is 61.10 cm (24"). It should be noted, however, that the maximum size recommended for a 3.2 mm (⅛") sheet of glass is 86.4 cm (34"). However, while plastic guidelines do not mention a critical length for the long dimension, there are limiting lengths for the glass long dimension; the 3.2 mm sheet should be a maximum of 86.4 × 194 cm (34 × 76").[30] Perhaps it was

necessary to drop to a glass dimension of 24" on one side in order to structurally carry distances greater than 76" on the other. Also, selection was probably based on glass sizes that are readily available, on their replacement cost in the event of breakage and on the tolerances required to accommodate thermal expansion. So, a 24" width is often used as a short dimension for both plastic and glass.

Plastic surfaces also scratch very easily, and this scratching can lower the overall transmissivity by scattering the light at the outer surface of the plate. This results in less light reaching the absorber.[31] Glass, on the other hand, is very resistant to scratching and therefore it would not be a problem with glass coverplates. If plastic sheets are necessary, there are special coatings or films that can be applied to the surface of the plastic that would make it scratch-resistant. But again, special coatings make the cost go up for the plastic coverplate.

Most of the comparisons so far have been between glass and plastic sheets. However, there are plastic films as well. The plastic films available have slightly different advantages and disadvantages from the sheets. Some of the more common films are: polyethylene, polyvinyl fluoride and fluorinated ethylenepropylene. The main disadvantage is that they have a short life under outdoor conditions. They degrade under ultraviolet light; they sag badly at elevated temperatures; they fatigue and tear under the constant flutter induced by wind stress; and continual exposure to moisture can also lead to failure, since water can often actually change the chemistry of the film, either by hydrolysis or by leaching out certain chemicals.[32]

The primary advantage to thin films is their considerably lower cost and the fact that the thin films (2 mil) will transmit up to 97% of the solar energy, where most glass and sheet plastics (3.2 mm, ⅛") operate at a reasonable high of 85-90% transmitance. However most plastic sheets and films are also transparent to reradiated heat (infra-red) with a transmissivity between 75-85%. "Tedlar" does reduce reradiation to 30% transmittance, but this is still a lot when compared to virtually 0% infra-red transmittance for glass.

Part of the "Greenhouse effect" is the fact that it traps the infra-red spectrum. That a plastic does not do a particularly good job of trapping the infra-red only serves to lower its effectiveness as a cover material. This greenhouse phenomenon is not too pronounced in low temperature, solar applications.

Since the main advantage in using thin films is the low initial cost, it is particularly appropriate for short-term solar installations. For a long term installation, with repeated replacement of the films, the added costs of material and labor to re-cover the collectors has to be considered and designed for. In the long run, glass may be cheaper. Efforts to produce a longer lived film narrows any economic advantage the films may have by making them more expensive.

As for general maintenance, dust build up and dirtiness have been found to be accountable for only a small reduction in transmissivity of the coverplate, generally 2-4%, and with any kind of angle or tilt, glass plates are relatively self-cleaning with occasional rains.[33] The plastics, though, seem to cling to the dirt a little more and it would take a more positive effort to thoroughly remove this build-up.[34]

As mentioned before, in all respects—except for its resistance to breakage—glass is superior to plastic materials; and, in addition, glass generally costs less than plastic sheets of the same thickness. The only reason to ever really consider plastic **sheets** is if breakage should be judged as a serious problem. The only reasons to ever consider plastic **films** is if the installation is short termed, or if the maintenance costs for complete replacement of the film every 2 to 5 years over the life of the collectors was still less than the cost of glass coverplates. One should keep in mind that the cheaper plastics available (close to the cost of glass) do not seem to have the properties one needs in a coverplate, and that for each

treatment—ultraviolet protection, infrared reflection, scratch resistance, water leaching—the cost must go up.

Recent literature has suggested that some hybrid plastic/glass laminates might be a solution. But these have their own attendant problems, such as having to select plastics with the same index of refraction as glass, and solving the problems created by the stress that is induced as the sheets expand at different rates when heated.[35] Once again, cost will be a very important factor.

The selection of the number of coverplates for a system will depend on the temperature and efficiency at which the collector is designed to operate. Efficiency is typically the measure by which collectors are judged, and it is generally defined as the amount of heat actually collected compared to the amount of solar radiation incident on the collector surface.

$$\text{Eff.} = \frac{(T_{out} - T_{in}) \ (\text{Fluid: Weight/Time} \times \text{Sp. Ht.})}{(\text{Insolation/Time} \times \text{Collector Area})}$$

Where: T_{out} is the outlet temperature of collector,

T_{in} is the inlet temperature of the collector and

Sp. Ht. is the specific heat of the collecting fluid.

When investigating the performance of a collector it is important to understand what the efficiencies quoted by a manufacturer really mean. Some manufacturers define efficiency differently from that above in order to make the performance of their collectors seem superior to others on the market.[36]

With the above efficiency in mind, Figure 3.30 compares the performance of collector assemblies with zero, one, two and three coverplates. As the temperature required of the absorber goes up, the pressure behind various modes of heat loss goes up;

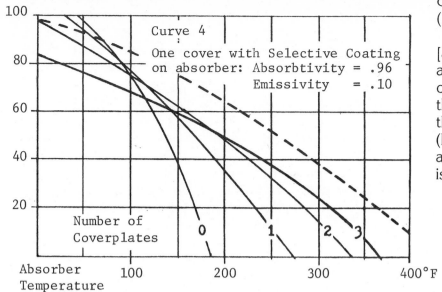

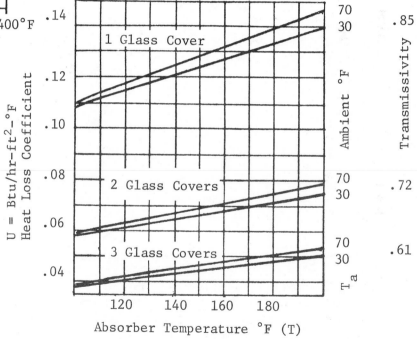

Curve 4

One cover with Selective Coating
on absorber: Absorbtivity = .96
Emissivity = .10

Number of
Coverplates

Note: - Absorber is painted flat black,
absorbtivity = .96, emissivity = .96.
- Ambient air temperature is 70°F
- Flow rate is adjusted to yield desired
absorber temperature and efficiency

[37]**Figure 3.30**

and, as heat loss goes up, efficiencies drop. Since one of the primary functions of coverplates is to prevent heat loss, collectors that have to operate at relatively high temperatures will have higher efficiencies when more coverplates are used to suppress heat loss (Figure 3.31). However, if a system can be designed to operate at lower temperatures, much better efficiencies are available and fewer coverplates can, and in some instance should, be used. Each coverplate used cuts down the energy reaching the absorber (Figure 3.31), and when collector temperatures are relatively low and there is less pressure on heat loss mechanisms, the loss of energy in the lower , transmissivity due to multiple

coverplates is not offset by the reduction in heat loss (Figure 3.32).

For most applications, moderate temperatures [around 60°C (140°F)] are more than adequate,[27] and two coverplates are generally all that is needed or wanted. With two plates, the heat losses are half of those with just one plate. The improvement due to the addition of a third plate is not quite so marked (Figure 3.31). When the added cost of the third plate and its poor low-end performance are considered, it is not a particularly advantageous alternative for col-

Q_L = Total Heat Loss through top
of collector: Btu/hr.
A = Area of collector: Ft².

$$Q_L = U(T - T_a)A$$

[38]**Figure 3.31**

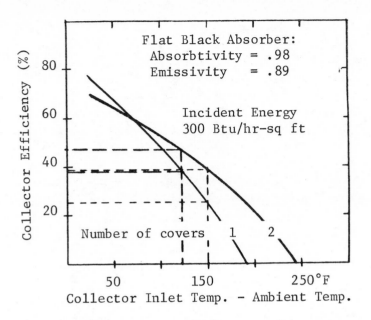

Flat Black Absorber:
Absorbtivity = .98
Emissivity = .89

Incident Energy
300 Btu/hr-sq ft

Number of covers   1   2

Collector Efficiency (%)

Collector Inlet Temp. - Ambient Temp.

[41]**Figure 3.32**

lectors operating at moderate temperatures.

In climates where winters are relatively mild—above an average temperature of 10°C (50°F) or so—one coverplate ought to suffice to produce the temperatures required for space heat and for hot water.

Three coverplates should be used only when high temperatures are required. The only likely candidate for such use is absorbtion refrigeration air conditioning which requires temperatures around 85-120°C (180-250°F).[40] However, once temperatures exceed 93°C (200°F), it would probably be more economic to chose an absorber with a selective coating that needs only one coverplate (Figure 3.30).

It should be noted that the efficiencies in Figure 3.30 are based on a 21°C (70°F) ambient air temperature. As the temperature outside the collector drops, the difference between the absorber temperature and air temperature increases and the efficiencies decrease. Figure 3.32 compares the performance of one and two coverplate assemblies as a function of the temperature difference between the absorber and ambient temperatures—for a constant solar flux. So, in the winter, if the outside air temperature is −18°C (0°F) and if hot water is required at 65°C (150°F), a single cover collector will operate at 23% efficiency and a two cover collector will operate at 38% efficiency—an improvement of 65%.

However, if hot water at 49°C (120°F) could be used instead of 65°C water, then the temperature difference would be less and both collectors would operate more efficiently. The single plate assembly would operate at 38% efficiency and the two plate assembly would operate around 48%. The two plate produces 25% more usable energy than the single plate device, but as the operating plate temperature gets lower, the improvement due to the second plate becomes marginal. According to Figure 3.32 the cross-over in efficiency occurs around a point 39°C (70°F) higher than ambient temperature and at an efficiency of 60%. Therefore, in climates with average winter temperatures around 10°C (50°F) hot water at 49°C is at a temperature 39°C above ambient, and all that would be needed is a collector assembly with a single coverplate. In fact, a two cover assembly would be wasted under these conditions. Although the cross-over occurs at a point 39°C above ambient and at 60% efficiency, the economic cross-over will occur farther down the curve at some point where the extra energy recovered would offset the cost of the added coverplate.

The figures above are based on a constant solar input of about 950 Watts/m² (300 Btu/hr-ft²), which is the high end of probable hourly insolation values. Figure 3.33 shows what efficiencies are available at given levels of insolation and at set collector temperatures. It illustrates that collector temperatures can be maintained at a set point even as the solar energy level varies: by adjusting the flow

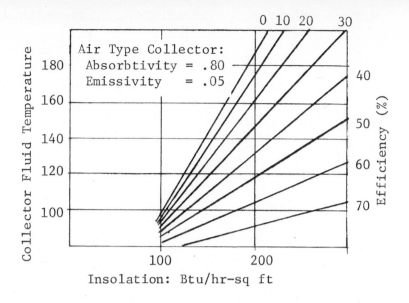

[42]**Figure 3.33**

If a design requires two or three coverplates, there are various combinations of glass, plastic sheets and plastic films. For two cover assemblies there are nine permutations, but only two of these are really acceptable for any long-term installation (Figure 3.34). Because plastics should not be used as interior covers in collector assemblies, the inner cover plate should always be glass. This maximizes the greenhouse effect, trapping the infrared radiation. It would also protect the outer, plastic plate from the high temperatures that could destroy it; and it would buffer the effect of expansion, since, if the temperature in the space between the glass and plastic plates is significantly lower than the temperature between the glass plate and the absorber, then the actual expansion of the plastic plate could be markedly reduced and might be more compatible with the tolerances one would design for a glass plate at its higher temperature. With the plastic plate outermost it functions as a breakage resistant cover—which is the only advantage it has over glass.

rate of the heat transfer media the temperature of the fluid can be manipulated, changing the efficiency but holding the set temperature. This accommodates the change in insolation.

Since the efficiencies for collectors can vary with the input, it will be up to the designer to decide where the economic optimum is for the performance standards set and to select the number of coverplates to be used on the collector to provide the efficiencies required.

> **CAUTION: The data used in Figures 3.32 and 3.33 are for specific collector designs. Each manufacturer should produce similar, independently tested curves for his particular product.**

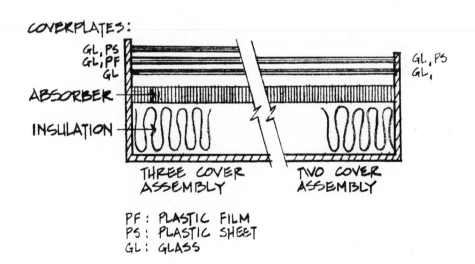

Figure 3.34

It would, in fact, seem that the only reasonable position for plastic plates is as an exterior cover in a high breakage zone.

If breakage is not expected to be a problem, then the second cover should also be glass. In Figure 3.35 the heat loss for several coverplate combinations is compared to their transmissivity and to absorber temperature. A plastic plate with a glass one would suffer from the poor transmission of the two glass plates and from the high heat loss of the plastic film over glass combination.

Thin plastic films, on the other hand, do not work very well as exterior covers. They need to be protected since they fatigue rapidly under wind flutter. And, since they deflect readily under impact loads, they are virtually useless in preventing the breakage of the interior glass plate.

For a three cover system, there are 27 possible combinations; but only four of them are acceptable (Figure 3.34). The inner cover should be glass and the outer cover can be glass or plastic sheets. The middle cover can be glass or plastic film; it should not be a plastic sheet because it would do no more than a film to reduce heat loss but it would cause a greater reduction in transmission.

If a plastic film is used as a middle cover, it is protected from deterioration on both sides and it would reduce transmissivity only slightly compared to a two cover assembly. The use of a film would help alleviate two of the major problems of the three cover systems: their severely low transmissivity and the expense of adding the third plate, since plastic films are relatively inexpensive.

As with any decision that will affect the overall cost of the solar installation, the selection of the number of coverplates used will have to be cost effective. It will be very important that added costs reap very good improvements in the total performance of the system.

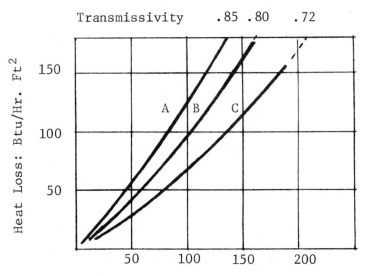

Key: A – One Glass Cover
     B – One Glass and One Plastic Film
     C – Two Glass Covers

[43]**Figure 3.35**

Absorber: The absorber is the most important part of an active collector. This does not mean, however, that the other elements are not essential as well. In discussing absorbers, the most important breakdown is by the kind of fluid it uses to transfer heat. The fluid will be either a liquid or a gas. In general, the liquid is water, or a water/antifreeze mixture; the gas has always been air. The absorbers fall into two general categories: air type and water type absorbers. Their relative advantages and disadvantages will be discussed.

The main function of the absorber is to capture the solar energy and transfer it to a working fluid. This requires the transfer of heat. As mentioned before, heat transfer is dependent on a number of variables:

Q = Heat Transfer
  = ΔT(Mass) (Time) (Specific Heat) (Surface Area) (Absorber Conductivity)

(ΔT is the average temperature difference between absorber and transfer fluid.)

Air Type Collectors

Air has a lower specific heat than water [.24 compared to 1.0 cal/(gram · °C) (BTU/lb. · °F)]. Because the properties of air are quite different from the properties of water, an air collector will be designed somewhat differently and will have different operating characteristics. With respect to water heating solar collectors, air heaters can be expected to move more mass, operate at higher temperature differences and utilize more surface area for direct heat transfer to the air: all because of the lower specific heat of air. To recover the same amount of heat (Q) an air system adjusts the parameters of the above equation:

Q = ↑ΔT (↑Mass)(Time)(↓Specific Heat)(↑Surface Area)k

Providing enough surface area is a concern in an air collector. Figures 3.36-3.41 illustrate several types of response to the problem of providing enough heat transfer area to the air.[44]

Figure 3.36 shows an absorber with fins on its back surface, like a radiator. As the air moves through the back of the collector, parallel with its fins, it cools them, absorbing the heat to itself by conduction. The front of the absorber plate is treated with black paint or a selective surface coating to enhance its absorption of energy. In this particular example, since the absorber has to be very effective at moving its heat out along the fins, it is important that the absorber and fins be fabricated of materials with high conductivities; this is in addition to the requirement that it have a large surface area for the conduction of the heat to the air. The combination of high conductivity and a lot of surface area dictates the use of large amounts of material—generally steel or aluminum. These materials are expensive, both in the drain on money and on resources; and they should be used sparingly.

Upon closer examination Figure 3.36 shows that the air made only one pass through the collector. Figure 3.37 shows a similar configuration, but the air makes two passes through the collector. This extra pass provides additional surface area for the air: more heat can be picked up by the air with very little additional effort. Making two passes maximizes the use of the surface area of the absorber plate.[45] It also

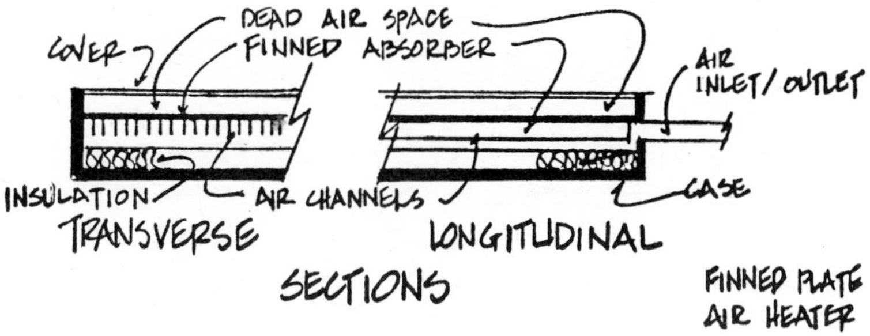

Figure 3.36

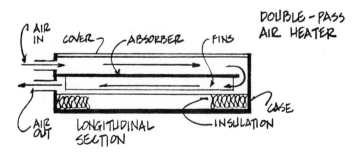

Figure 3.37

represents a savings in the amount of energy needed to run fans. The double pass means more heat can be collected by the same number of cubic feet of air—at virtually the same power requirements for the fans. This is primarily because the **time** factor for heat transfer has been increased (doubled). However, keep in mind that a double pass will not allow one to reduce the amount of collector area needed, since the amount of energy that can be collected depends on the level of solar radiation and the number of square feet one spreads out to intercept this radiation.

The significance of the double pass is that it reduces the amount of energy expended just to collect the heat for later use. This reduction enhances the positive side of an overall energy balance which has to deduct the energy needed to run pumps or fans from the amount of usable energy collected. Such a balance should be done to ensure that significantly more energy is stored than is consumed just to collect it.[46]

Figure 3.38 shows an air collector referred to by Yellot as a "Matrix" type. The typical absorber material in this case is a black gauze. It is black for high absorbtivity and it is a gauze (or something similar) for its porosity with respect to air. The fibrous nature of the material provides a great deal of surface area for heat conduction to the air. The principle considerations in selecting a material of this type are its "stiffness", and its tolerance to heat and light.

For stiffness, it should resist any tendency to pack down, closing its pores and creating a great deal of resistance to the passage of air through the gauze. Choosing a material that is resistant to moisture would probably be the first step towards insuring that the absorber maintains its stiffness, since water tends to make fibrous materials mat together. For heat and light, it should be as carefully chosen as any plastic coverplate with respect to its useful service temperature and its resistance to ultraviolet degradation.

Because of the interwoven nature of the material and because it does offer a great deal more surface area than simple metal fins, there is a higher coefficient of friction which works to impede the passage of air. This impedance, or back pressure, will have to be determined for each matrix type absorber. Appropriate measures have to be taken to insure that there is a balance between necessary surface area for effective heat transfer and the amount of energy needed to power the fans that have to overcome the back pressure.

In general, absorbers have been fabricated from metal sheets. Both the material and the fabrication process are expensive. In fact, the absorber represents the single most expensive item in the collector; and the collector is the most expensive part of a solar heating/cooling system. This cost, however, should be greatly reduced with matrix type collectors, since black gauze is invariably cheaper than metal plates.

Figure 3.39 represents a design by George Lof for an air collector system that has been in operation on his home for quite some time. His scheme utilizes glass plates as absorbers and as fins to direct the air and to provide sufficient surface area for heat transfer. One half of the surface of each plate has been painted black to absorb the solar energy, the other half is clear and laps the blackened surface of

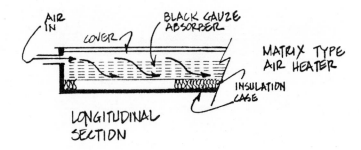

Figure 3.38

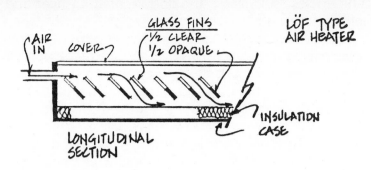

Figure 3.39

another plate. This arrangement allows sun to shine through to each blackened absorber surface and the glass fins channel the heat absorbing air between two blackened glass sections, one above and one below the air stream.

Another type of air heating collector, the vee-corrugated (Figure 3.40), was developed by Australian Scientists of the Commonwealth Scientific and Industrial Research Organization.[47] The parallel vee-shaped grooves reflect the sunlight several times down the "V". Less light is allowed to reflect away from the surface and, in this respect, it is markedly better than a normal flatplate absorber. However, this corrugated absorber requires at least twice the amount of metal plate compared to the amount of metal needed for a flatplate absorber of the same size. Nevertheless, the better performance might justify it. This is one area that should be the subject of further investigation. The better performance will have to be weighed against the higher costs. Another advantage of the "V" type collector is that its corrugations provide a good deal of surface area for thermal conduction to the air, which flows in the trough beneath the vees. This system would be well suited to an installation which has to have limited collector area.

An air collector operating on a principle similar to the "V" grooved collector has been developed by John Keyes. Figure 3.41 illustrates his "beer can" collector. The walls of the beer cans are painted black and they work to absorb light as well as to reflect it down into the cups. Air is blown over the cups at high velocity, creating turbulence and effecting better heat transfer in the cups and sucking heat out of the cups. This system may operate at fairly good efficiencies, but it requires a large amount of material (generally aluminum) for the absorber. This system is coupled with reflective insulating panels like the Baer drumwall (Figure 3.16), which can make the collector performance markedly better.

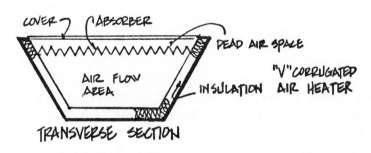

Figure 3.40

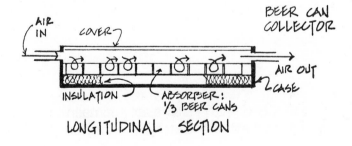

Figure 3.41

A corporation called **Solaron**, markets an air collector that has a simple black air duct (Figure 3.42) for an absorber. Air collectors have been designed by LPM as ducts that can be fabricated by any sheet metal shop, and installed by carpenters and glaziers. Using basic duct work simplifies the design considerably and effects lower costs.

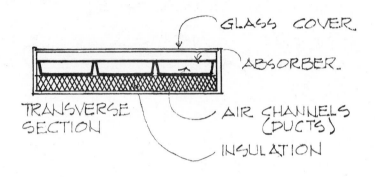

Figure 3.42

Fluid Flow Rates: The amount of air to be circulated through a solar collector is very much a function of the collector design and the temperatures it needs to deliver to load. Doubling the flow rate will have the effect of cutting the temperature difference between inlet air temperature and outlet air temperature by almost one-half. Conversely, cutting the air flow by one-half will roughly double the ΔT between inlet and outlet. In general, it has been established that 8°C (15°F) is a reasonable maximum ΔT for which to design.[48] Based on recommended flow rates for water in the 1978 ASHRAE Handbook of Applications,[49] comparable flow rates for air would be between 0.5-1.5 m³ per minute for each m² of collector area [1.8-4.8 ft³/(ft² · min) of collector area]. The higher end of this range will keep the ΔT between the collector and ambient air lower, raising

efficiencies. D. J. Close has documented relatively high efficiencies with 1.4 m³/(m² · min) (4.5 ft³/(ft² · min). Kreider and Kreith suggest that there are only marginal gains with flow rates above 0.94 m³/(m² · min) (3ft³/(ft² · min)).[50]

There are some basic advantages to the air heating collector. There is no need to worry about leaking. Corrosion of the absorber is no problem. Equipment failure (valves vs. dampers) is non-critical. If there are leaks, there is no loss of expensive chemical treatments for the water. There are no maintenance checks to make sure chemical levels in the water are appropriate.

Air collectors will also "survive" at the lowest temperature a collector may encounter without the fluid ever freezing, so there is no danger of pipes and plates ever rupturing, nor is there a need to "drain" the system—all of which are considerations in a water system.

Air systems also have the advantage of being directly compatible with forced air heating systems. Solar energy can be delivered straight to load without intervening heat exchangers. This helps to improve system efficiencies.

Water Type Collectors

The following are some general considerations that apply to all liquid heating collectors. Each system has to be designed so that it will not leak. The freezing point of the liquid has to be low enough so that it will not freeze in the collectors under the most severe cold weather one would expect; or the system must be designed so that the collectors will drain themselves whenever they are shut down. The compatability between the liquid used to transfer heat and the various mechanical parts in the system has to be investigated. This is to insure that there will not be any unexpected corrosion problems. Mixing and matching different metals for the various components should be done with great care. Aluminum, steel and copper are the most widely used materials; and, in the presence of water, aluminum undergoes galvanic corrosion if it is in contact with copper or iron. So it should not be used with copper or galvanized pipes. There are things that can be done to inhibit this corrosion, but for each problem the design creates, the solution just makes the whole package cost more.

It would be best not to mix materials in the system. Copper pipes should be used with copper absorbers, galvanized pipes with steel ones and, with aluminum absorber systems, it would probably be best to use plastic pipe, which would be inert as far as galvanic corrosion of the aluminum is concerned. There are different kinds of plastic pipe, however, and the ones that would be able to resist the heat to be expected from collectors are quite expensive.

The costs of collector systems are very tightly tied to the costs of the absorber material. Copper is the best material for conducting heat across or through the absorber plate to the fluid. Copper has a conductivity—usually "k" in equations—of 98 calories/second · meter · °Celsius (Cal/(sec · m · °C)) (Figure 3.43),[51] but copper is the most expensive of the three common metal absorbers. Aluminum is next in performance and in price; it has a conductivity, k, of 48 cal/(sec · m · °C). A steel absorber plate has a conductivity, k, of about 11 cal/(sec · m · °C).

Aluminum, however, has an environmental disadvantage in the fact that it requires more than 8 times as much energy to make aluminum as it does to make steel; 64,000 BTU/lb for aluminum compared to 7,500 BTU/lb for steel.[52] Copper is not much better; it requires 42,000 BTU/lb. Unless new ways to produce aluminum can be found, it might be difficult to justify the production of an energy saving device when it is made from such energy intensive material.

The conductivities are indicators of how readily a material will move heat over distances; and, in the past, they have been indicators of how costly the fabrication of a collector might be. A common approach to constructing FP collectors has been to take a sheet of metal—an absorber—and clamp or

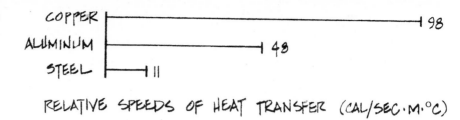

RELATIVE SPEEDS OF HEAT TRANSFER (CAL/SEC·M·°C)

Figure 3.43

bond (Figure 3.44) water carrying tubes to the surface of the absorber. The absorber's conductivity is very important since it has to transport the heat to the tubes (Figure 3.23). If a material has a low conductivity, the tubes have to be spaced closer together to perform as efficiently as an assembly made of material with higher conductivity and wider tube spacing. Consequently, more tubes are needed and this drives up the cost of materials and labor—since both are based on the number of feet of tubing that have to be bonded to the plate. However, in an article by Bliss, it was shown that there was little or no dif-

TUBE-ON-PLATE
FABRICATION PROCESSES

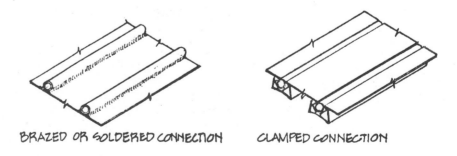

BRAZED OR SOLDERED CONNECTION   CLAMPED CONNECTION

Figure 3.44

ference in collector performance, due to different absorber materials, if the tubes are spaced between one and two inches on center.[53] This would be essentially equivalent to a continuous bundle of tubes.

It is the higher conductivities of copper and aluminum that makes collectors of these two materials more attractive than steel collectors, since the higher cost of these more expensive materials could be offset by requiring less material and fabrication. However, recent methods of absorber plate fabrication have done much to lower fabrication costs for aluminum and steel. For aluminum and steel there is the "roll bond" method (Figure 3.45) where two sheets that have the pattern of tubes laid out in graphite between them are pressure bonded to one another. In the process, the plates do not bond—or fuse—to one another wherever the graphite pattern is. The fused plates can then be inflated and tubes, following the graphite pattern, "pop" up under the pressure. The rest of the plate stays flat since it has been, essentially, pressed into one sheet. Copper has a similar technique which uses ingots instead of sheets.[54]

These roll bonded absorber plates have made substantial inroads into the current collector industry and are replacing the older processes of clamping or brazing tubes on absorber sheets.

For steel, an additional fabrication technique is available.[55] Two sheets of steel are fabricated together with a continuous edge weld and a matrix of spot welds over the surface (Figure 3.46) of the absorber plate. The sheets, which can be fabricated as a continuous strip, are then pressurized and the whole surface forms a matrix of little "dimples" (Figure 3.47). This gives the steel sheet a quilted appearance. But, more importantly, it also simulates the characteristics of an absorber made from a continuous bundle of tubes, which according to Bliss' figures (Figure 3.48) makes its performance equal to the best copper or aluminum systems. This method eliminates the extra materials for tubes and the extra labor for securing the tubes to the plate. This fabrication process should markedly enhance the attractiveness of steel collectors, by lowering the square foot costs substantially. Another advantage of this steel collector is that, as previously discussed, it can be used structurally as the roof.

There is one significant disadvantage to the steel collector: it is subject to corrosion (rust). This problem has to be solved with respect to the internal fluids as well as the external coatings. The external surface will probably be painted or coated in order to optimize absorbtivity, so this process could be coupled with efforts to prevent the oxidation (rusting) of the treated surface.

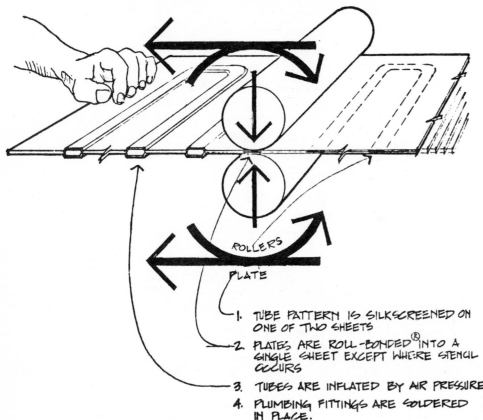

1. TUBE PATTERN IS SILKSCREENED ON ONE OF TWO SHEETS
2. PLATES ARE ROLL-BONDED® INTO A SINGLE SHEET EXCEPT WHERE STENCIL OCCURS
3. TUBES ARE INFLATED BY AIR PRESSURE
4. PLUMBING FITTINGS ARE SOLDERED IN PLACE.

Figure 3.45

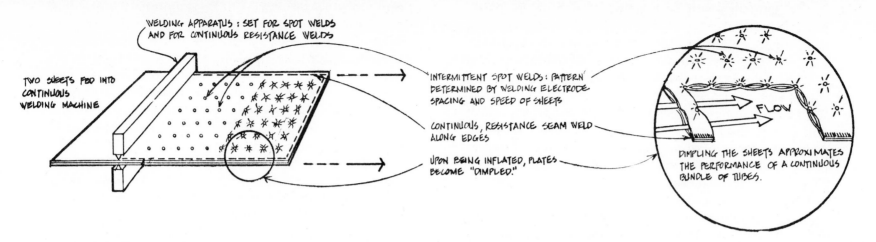

WELDING APPARATUS: SET FOR SPOT WELDS AND FOR CONTINUOUS RESISTANCE WELDS

TWO SHEETS FED INTO CONTINUOUS WELDING MACHINE

INTERMITTENT SPOT WELDS: PATTERN DETERMINED BY WELDING ELECTRODE SPACING AND SPEED OF SHEETS

CONTINUOUS, RESISTANCE SEAM WELD ALONG EDGES

UPON BEING INFLATED, PLATES BECOME "DIMPLED."

FLOW

DIMPLING THE SHEETS APPROXIMATES THE PERFORMANCE OF A CONTINUOUS BUNDLE OF TUBES.

Figure 3.46

Figure 3.47

In order to prevent the collector from rusting from the inside out, it has been suggested that special fluids be used to neutralize the tendency to rust. These special fluids—just like every speciality item in collectors—can be very expensive. Water can be used if it has been chemically treated with rust inhibitors. However, the balance of these chemicals has to be checked occasionally to ensure that the inhibitors are kept up to strength—similar to checking the acid in a battery.[57]

Other fluids tried have more exotic chemical bases such as silicon, polygol, glycol and various oils. They have different properties from water and from each other. Their viscosity is generally greater than that of water and more pump power would be required to circulate the fluid. Their specific heat is generally lower than that of water, so more pounds of the fluid would have to be pumped to collect the same amount of heat. Besides the expenses of these fluids, some of them are toxic.[58] For this reason they operate in what is called a closed cycle, which is discussed in the chapter on storage. A closed system insures that these fluids are not just pumped through the collector and down a drain somewhere or into the house hot water supply.

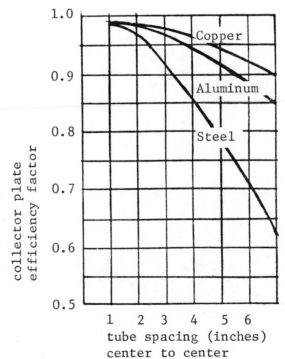

CONDITIONS:
varies

1. absorber-tube configuration

2. tube diameter – ½"

3. turbulent flow

4. loss rate coefficient for collector = 2.5 Btu/sq.ft.-hr-°F

Note: 1" spacing approximates a continuous tube bundle or a plate that is one large, narrow tube. Examples:

[56]**Figure 3.48**

The steel collector referred to is made from "mild" or carbon steel. This is the common variety of steel that requires steps be taken to prevent rust. If none of the above solutions for prevention of rust are shown to be satisfactory over a long period of time; or if the possibility of corrosion **must** be avoided, then a different absorber material has to be used. With aluminum, there is the potential problem of galvanic corrosion due to dissimilar metals. If not properly prevented, galvanic corrosion can be much worse than rust; so aluminum, though generally more expensive than steel, will still not provide protection against absorber degradation due to corrosion. Stepping up in price to copper absorbers is probably the most reasonable solution to avoiding corrosion problems. As long as the working, heat transfer fluids do not specifically attack copper, the absorber is likely to last 20 plus years. Stainless steel is another option that is available, but it is prohibitively costly in all but the most corrosive of operating environments.

Selective Surfaces: The purpose of selective surfaces in solar technology is to enhance certain performance characteristics of the absorber. The idea is to develop a surface that will have the highest possible absorbtivity and the lowest possible emissivity. A simple coat of black paint will increase the absorbtivity of a plate, but emissivity is controlled by insuring that the surface of the plate under the paint remains highly polished—mirrorlike. This may or may not be a hard thing to achieve at reasonable cost. However, if the collector is fairly efficient, and it does not have to operate at the high end of the FPC range—95°C (200°F), then radiative heat loss (emissivity) should not be too severe a source of heat loss. In this case, the use of black paint to enhance absorbtivity, with no special effort to reduce emissivity, should be sufficient treatment.

There has been a great deal of attention given to the control of emissivity in the literature. This probably reflects the desire of scientists and technicians to develop high performance FP collectors for high temperature applications. These criteria **create** the problems that selectively engineered surfaces are designed to solve.

The higher the temperature that one wants to extract, the higher the losses one has to counteract. Conduction losses, for example, are directly proportional to the temperature difference between the transmitting and receiving objects. Consequently, attempts to get process heat at 93-100°C (200-212°F) when the surroundings are 4-27°C (40-80°F) creates a temperature difference, ΔT, in the neighborhood of 85°C (158°F). If the collectors were used to produce hot water for domestic use, they would operate best around 71°C (160°F); this would be an average temperature difference of 55°C (100°F). This means that under the same conditions, the collector used to produce higher temperatures would lose at least 50% more heat through conduction than the collector that is used to produce lower temperatures. These are heat loss figures for conduction only. There are additional heat losses due to radiation which also are temperature dependent. But radiation losses increase according to the fourth power of the temperature difference (ΔT).

$$\text{RADIATIVE HEAT LOSS} = K(T^4 - T_0^4)$$

where K is the Stephen-Boltzmann constant, T is the absolute temperature of the absorber, and T_0 is the absolute temperature of the surroundings.[59]

Because of this fourth power effect, control of radiative heat loss with selective surfaces becomes very important as the operating temperature of the collector is forced higher.

The discussion of flatplate collectors has proceeded through gradually more complicated problems and solutions involved in the utilization of FPCs. Most of the problems have to do with heat loss and its prevention. This problem gets more and more sophisticated as technology pushes the FPC up to higher and higher plate temperatures. To do this it is necessary to try to isolate the plate from its surroundings. The first step is to cut off convection with the outside air by the use of coverplates. The second step is to isolate the plate from contact with the casing as much as possible in order to cut down on possible paths of heat conduction to the outside—insulation is an important part of step two. These elementary steps have a significant effect in raising the temperature of a basic, black absorber plate. However, now that there has been an effort to raise the temperature of the plate, it is found that the heat losses have also risen and that the greatest source of loss is now the radiation from the heated absorber plate.

Reducing the radiation is accomplished by reducing the emissivity of the plate; the lower the emissivity the less the plate tries to radiate the heat away. Low emissivity is a function of the surface of a material. Shiny, reflective, highly polished surfaces have very low emissivities; however, they also have very low absorbtivities. On the other hand black or rough surfaces have very high absorbtivities, and very high emissivities. A selective surface is created by depositing or electroplating a black surface of specific thickness on a highly polished, mirror-like metal base. This keeps the properties of each separate and intact.

The common solution is to take a polished absorber and put a very thin black surface, generally a metallic oxide coating, on top. The contact of this thin, black oxide coating (which has very high absorbtivity) with the highly polished absorber (which has very low emissivity) makes sure that the heat and light absorbed at the surface of the black coating is quickly passed through to the metal plate by conduction. And, once the plate heats up, the polished surface inhibits radiation from the plate. In this way the best properties of the plate and its coating are melded to produce a significant change in the performance of a collector absorber plate. At this point, however, it is likely that the marginal improvements in the performance of flatplate collectors has passed the point of diminishing returns; the added com-

plexity and cost for relatively small increments in improved performance are not justified economically.

There is one more measure in heat loss prevention that is significant because it has received serious treatment from several sources in academia and in business. It is also significant because rising fuel costs may ulitmately make every measure to save energy economical. The extra "twist" comes with building a collector that has a completely sealed housing capable of sustaining a moderate vacuum. It was found that, after going to the trouble of designing an absorber with high absorbtivity and low emissivity, the most significant mechanism for heat loss was convection from the absorber plate to the coverplate. The process of convection depends on the movement of air, so the solution to this convection problem is to remove the air.

Figure 3.49 shows a cross section of a vacuum flatplate collector. According to a paper at the ISES conference by C. A. Eaton and W. A. Blum, small studs are structurally necessary to resist the pull of the vacuum on the two molded plexiglass pans. These studs account for 3-4 percent of the surface area of the collector. The actual absorber plate is

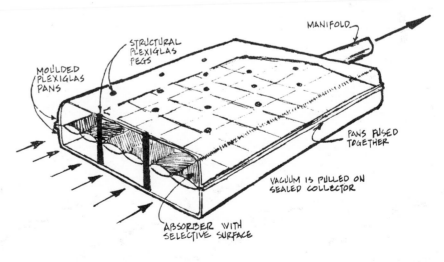

Figure 3.49

suspended between the two sheets of plexiglas; the only absorber contacts with the casing are the inlet and outlet pipes.[60]

A recent series of tests conducted by NASA shows the relative losses of a flatplate collector with a selective surface.[61] The plate had an absorbtivity of .90 and an emissivity of .08. The test conditions were described as quasi-steady state. Figure 3.50 A and 3.50 B are schematic diagrams of the quantity of heat flow by different mechanisms during the tests.

Figure 3.50 A is for a collector at 74°C (165°F), which is a moderate operating temperature for a collector. In this case, it is seen that the convective heat losses are greater than the total of all the other losses combined, and that it is 6 times greater than the radiation losses from the selectively coated absorber.

Figure 3.50 B shows the same collector operating at a temperature of 138°C (280°F). Here, the convective losses between absorber and cover are extremely large compared to the total—2.5 times larger. These convective losses are what is being eliminated by the use of vacuum flatplate collectors. The efficiencies noted in Figure 3.50 are for comparative purposes only. NASA calculated 45% and −1% for their non-vacuum collector. These efficiencies were then recalculated on the assumption that the convective losses can be totally suppressed and that all other loss factors would remain the same. The new figures show efficiencies of 78% and 55%, respectively. The assumptions will not precisely model what actually happens, but they do serve to illustrate the fact that convective losses can greatly effect the performance of collectors that have already gone as far as trying to eliminate radiative losses by using a selective surface.

Note, again, that the higher the operating temperature of the plate is, the greater the heat loss will be. The losses at 138°C are significantly higher than those at 74°C.

Since these "fix-up" solutions are a response to the development of high performance collectors, it is understood that the plates have to reach high temperatures and they have to do it efficiently. Eaton and Blum cite a collector which operates around 115°C (240°F) at normal atmospheric pressure; however, if it is operated under a moderate vacuum it reaches temperatures around 180°C (355°F).[63] This represents a significant improvement in the performance of the FPC, but it is gained only after much fancy footwork and at a much higher cost than earlier the FPC discussed. Although these vacuum FPCs

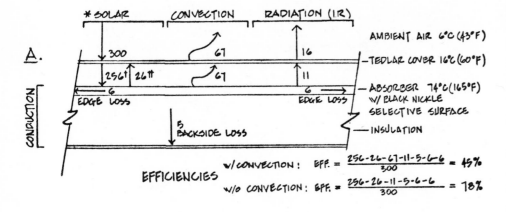

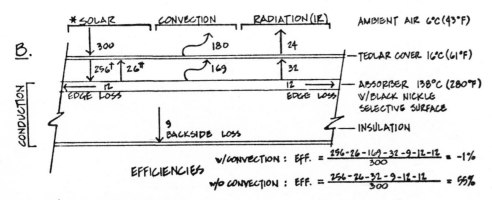

*ALL NUMBERS: BTU/H/① COLLECTOR
†.85 TRANSMITTANCE
‡.10 REFLECTANCE
(.90 ABSORPTIVITY)

[62]**Figure 3.50**

78

and other high performance collectors may be too high powered and too costly to be used appropriately in residential applications, there is probably a large market for them in providing high temperature hot water, in large quantities, for industrial uses.

It should be noted that the 115°C-180°C (240°F-355°F) temperatures that were just mentioned are **not** the operating temperatures of FPC. These are stagnation temperatures: no coolant is circulating through the plates. Under actual load conditions, with the coolant flowing, it is unlikely that even the vacuum, flatplate collector would exceed temperatures of 115°C (240°F).

If the collector runs at temperatures above the boiling point of the coolant, the system has to be pressurized to keep the fluid from boiling. If higher temperatures than this are needed, which will be beyond any practical flatplate collector, then it would be necessary to move into the realm of concentrating collectors to satisfy the higher temperature requirements.

In conclusion, technologically sophisticated collector devices find appropriate application primarily in commercial and industrial uses where there is a year-round need for relatively high-grade heat in large quantities. In such a case, whatever heat is produced can be fully used. For residential use, however, it would be a case of overkill to install such high-tech devices for winter space heating. Since economics is the most important factor in solar penetration of the residential sector, it would seem that the development of good air-heating collectors is most advantageous because they hold promise for being less expensive to manufacture. In addition, most residential units (66%)[64] already have ducted air heating, and solar retrofit will have to be compatible with existing heating units. A big boost to solar air heating would be the development of a reasonably efficient mechanism for heating water from hot air systems. This development would put a combined solar hot water and space heating system within the province of air collectors; and, thereby, improve the overall cost-effectiveness of the basic system. **Solaron** has such a package, but they are still in the general price range of water-type FPC systems.

PHOTOVOLTAICS

The idea that solar energy can be converted directly into electricity is a very attractive one. This is what a photovoltaic device does.

The actual device is generally a very thin "cookie" —around .16cm (.063 inches).[65] It is a lamination of two thinner "wafers" which are made of materials that have very different photochemical characteristics, and a "filling" that separates the two wafers. The conversion of light to electricity occurs when the electrons in the crystals of one wafer are so stimulated by the energy of incoming light that they "jump ship"; that is, they leave the particular atom in a crystal to which they had been attached and migrate across a buffer zone (the filling) into "slots" or electron-seeking holes in the other wafer. With this electron migration, the two wafers can be connected through a circuit and they will produce a current.[66] Now the current produced by each cell is very small, but when large numbers of these small cells (typically 4.4 cm² (.68 in²) are hooked up in a number of series and parallel circuits, useful amounts of current can be extracted from the system.

There are two common types of this form of photovoltaic device. They are the silicon cell and the cadmium sulfide cell. In general, the silicon cell has higher efficiencies and greater durability; it is also more expensive. However, even the relatively inexpensive photovoltaic devices have been much too expensive for general use. Photovoltaic devices have only been economically justified for very remote installations which have low power demands—as far as terrestial applications go. However, there are increasing numbers of research efforts to find ways of producing these devices at much lower cost. They may become a significant source of electricity in the future.

The better silicon—gallium arsenide cells

(Si/GaAs) are reported to be 13.5 percent efficient. This means that 13.5 percent of the energy that was in the sunlight was converted to electricity. It also means that 86.5 percent has to be dealt with as waste heat from the solar cells. There is some prospect that in the near future the efficiencies can be raised to 16-17 percent.[67] (There is a maximum theoretical efficiency in the neighborhood of 23 percent.[68]) However, the more efficient cells have definitely been more expensive as well.

Cadmium sulfide—copper sulfide (CdS/Cu$_2$S) cells are in operation at **Solar One**, a test house built by the Institute of Energy Conversion, University of Delaware, Newark, Delaware. These cells operate at an efficiency of 2.5 percent. Additional research since the installation of these cells has produced a CdS/Cu$_2$S cell with efficiencies as high as 8.3 percent. The expectation is that a stable cell can be produced with a continuous output of around 6 percent. In this case, approximately 94 percent of the intercepted sunlight becomes waste heat.[69]

With the conversion of most of the sunlight into heat, cooling the solar cells becomes a very necessary part of the design of a photovoltaic installation. The **Solar One** experiment designed the photovoltaic arrays so that they also operated as flatplate collectors. The collector design was very similar to that in Figure 3.51; the solar cells were glued to the absorber surface. As the cells heated up, they conducted the heat to the absorber plate which distributed the heat along its finned back. The heat was carried away by forced air circulation. This "waste" heat was then stored for use later in meeting the house heating demands. If the heat had not been removed, the cells would have degraded very rapidly.

In view of the fact that these are electrical devices, it would seem imperative that air be used as the coolant instead of water. Obviously, this would be best integrated with air-cooled flatplate collector systems. If a designer were to install solar heating, cooling and hot water systems, with the hope that he

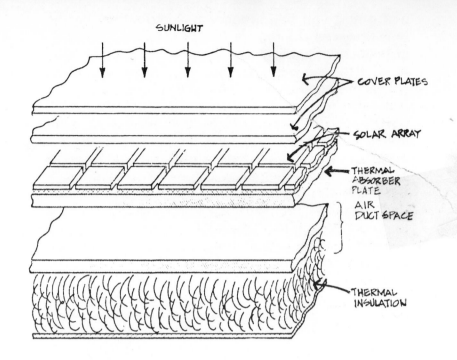

SUNLIGHT

COVER PLATES

SOLAR ARRAY

THERMAL ABSORBER PLATE

AIR DUCT SPACE

THERMAL INSULATION

[70]**Figure 3.51**

could later expand with photovoltaic arrays, once they become economically feasible (and they inevitably will), then he would probably fare best if he went with hot air systems for heat collection.

With **Solar One** the solar cells were sealed away from the circulating air to prevent dust build up, and they were hung beneath a skylight skin to protect them from the weather. Solar cells that are exposed to wind, rain and dust do suffer from surface deterioration. This can cause serious loss of power over time, as well as shorten the useful life of the cell. And, the life expectancy of a cell is a critical economic factor. An exposed cell might last five years, where as a protected one could be serviceable for 15 to 20 years. Obviously, the longer they last, the better.

A serious environmental problem that must be carefully guarded against is the hazard of heavy metal contamination. Cadmium is very toxic in sufficient dosages (it is in the same family as mercury) and it is an important component in Cd/Cu$_2$S cells. Arsenic is the problem in Si/GaAs cells. There could be considerable environmental damage done if no thought is given to the refining, fabricating, and ultimate disposal of such devices with these toxic materials. Perhaps the quantities involved in solar cells is inconsequential, but this should at least be determined before there is any unchecked proliferation of these metals in the environment.

The electrical power from solar cells is in the form of direct current (D.C.); power from utilities is alternating current (A.C.). If the solar cells cannot supply 100% of the electric power needs, then the dwelling has to also be tied into the utility network. This means that there have to be compromises. The DC power can be changed to AC power through inverters and rectifiers; but there is some loss of power in the process—very expensive losses when system efficiencies are already low. However, a large number of home electrical uses can be converted to DC operation; but, since motors usually require AC power, a change to DC would require the installation of two, parallel electrical systems. This would add to the overall cost of the solar power usage.

One thing to be wary of is not to be seduced into using the solar electric power to heat the dwelling. It has become electricity only after much effort and expense, and at very low efficiencies. This use might be attractive because DC power can readily be used in electric resistance heaters. But, this electricity is a very high grade form of energy and it should not be used to satisfy low grade energy needs if it can be avoided. These energy needs can, and should be, easily met by the heat collected while cooling the photovoltaic absorber panels.

In order to get the optimum use out of a solar electric system there has to be some means of storing the electricity to even out the differences between supply and demand. Curve "A" in Figure 3.52 shows the

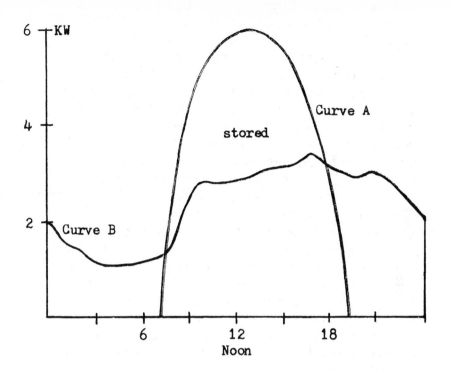

[71]Figure 3.52

relative intensities and the time of day that power might be collected. Curve "B" shows the power demand of an "average" house by intensity and time. The part of Curve A that exceeds Curve B should be stored for those times when Curve B exceeds Curve A. The most straightforward method is to use the solar electric energy to charge large banks of batteries. These batteries can be connected in arrays to match the current and voltage levels needed for use with the equipment in the house. In fact, it is a good idea to move solar power through the batteries before it is used; this would be done in order to level out the energy peaks from the cells. If the power were tapped directly from the cells, then there might be ups and downs in the power supply as clouds pass over. Going through the batteries would keep the supply even: pulling up extra power when a cloud passes by, and absorbing the surge of power when the sun pops out again. These batteries will be fairly expensive and they do have limited life times. Golf-cart type batteries can be expected to last about five years.[72] This area—electrical energy storage—is one that needs a lot of work. Current types of batteries seem to have been pushed to the limits of their capabilities. Electrolyzer/Fuel Cell technology is coming along, which will allow electrical energy to be stored indefinitely as hydrogen and oxygen gases; but it is a new technology that still needs much development to solve problems and bring down cost.

The technologies for utilizing solar electric power are available. They need more refinement and more incentives before manufacturers will commit whole industries to the level of mass production necessary to bring the price down and have some impact on the electrical demand of this country. Such a massive effort will probably require greater stimulation from federal sources, especially if solar-electric power is expected to ultimately supplant current sources of electric power.

Of the different proposed high-tech solar uses—both thermal and electric—photovoltaics is probably the only one for which there will be reasonably cost-effective solutions in the future. Direct solar thermal and thermal electric technologies seem to be at the technological limits for the components; photovoltaics, however, are an incipient technology that has not yet reached its limits. In fact recent developments in amorphous silicon cells (different from present crystalline cells) promises to lead the way to low cost solar power from photovoltaics.[73]

NOTES

1. Definition provided by the New Mexico Solar Energy Association through their slide program.

2. Paraphrased from Xenophon's Memorabilia, Bk. III, Ch. VIII.

3. Geo. F. Kech, ½ Utilities.

4. F. W. Hutchinson, "The Solar House: Analysis and Research," **Progressive Architecture**.

5. Victor Olgay, **Design with Climate**, p. 116.

6. **Ibid**., p. 117.

7. B. A. Peavy, F. J. Powell, and D. M. Burch, **Dynamic Thermal Performance of an Expermental Masonry Building**, Building Science Series 45, U.S. Department of Commerce, National Bureau of Standards, July, 1973, 99 pages. U.S.G.P.O. SD CAT. No. C 13.29/2:45 ($1.25). Stock Number 0303-01119.

8. J. Barnes, B. Lord, D. Mullman, and J. Stokoe, **Wall Design Handbook** (St. Louis: School of Architecture, Washington University, 1974), p. 26.

9. J. M. Fitch, **American Building: Forces that Shape It** (Boston: Houghton Mifflin Company, 1972), p. 269.

10. Jim Lackie, et al. **Other Homes and Garbage** (San Francisco, Sierra Club Books, 1975), p. 143.

11. Department of Housing and Urban Development, Office of International Affairs, "Information Series 24" (August 27, 1973), p. 4.

12. Harold Hay, "New Roofs for Hot Dry Regions," **Ekistics**, Vol. 183 (Feb. 1971), pp. 158-64.

    H. R. Hay and J. I. Yellot, "A Naturally Air-Conditioned Building," **Mechanical Engineering** (Jan. 1970), pp. 19-25.

13. Steve Baer, "Solar House," **Alternative Sources of Energy** (Kingston, N.Y.: Alternative Sources of Energy, 1974), pp. 39-41.

14. Bruce Anderson with Michael Riordan, **The Solar Home Book** (Harrisville, New Hampshire: Cheshire Books, 1976), pp. 108-109.

15. AIA Research Corporation for the Department of Housing and Urban Development, **Solar Dwelling Design Concepts**, (U.S. Government Printing Office, 1976), p. 14.

16. Paul Davis, "To Air is Human: Some Humanistic Principles in the Design of Thermosiphon Air Heaters," **Proceedings of the Passive Solar Heating and Cooling Conference**, May 18-19, 1976, pp. 40-45.

17. Bill Yanda and Rick Fisher, **The Food and Heat Producing Solar Greenhouse, Design, Construction and Operation** (Sante Fe, John Muir Publications, 1976), 162 pages.

    James C. McCullagh, Editor, **The Solar Greenhouse Book**, (Emmans, Pa., Rodale Press, Inc., 1978), 328 pages.

18. Londe-Parker-Michels, Inc., **Final Task Report** on Residential Redesign, Phase II of Building Energy Performance Standards, subcontract report to the AIA/Research Corporaton, Sept. 25, 1978.

19. Londe-Parker-Michels, Inc., "Early Field Experience and Cost Effectiveness of a Passive Solar Heating Retrofit," Proceedings of the 3rd National Conference on Technology for Energy Conservation, Tucson, Arizona, January 23-25, 1979.

20. J. I. Yellot, "Utilization of Sun and Sky Radiation for Heating and Cooling of Buildings," **ASHRAE Journal** (Dec. 1973), pp. 31-42.

21. S. W. Moore, J. D. Balcomb, and J. C. Hedstrom, "Design and Testing of a Structurally Integrated Steel Solar Collector Unit Based on Expanded Flat Metal Plates," International Solar Energy Society, U. S. Section Annual Meeting (Aug. 1974), p. 6.

22. Clarence W. Clarkson and James S. Herbert, "Transparent Glazing Media for Solar Energy Collectors," ISES Conference, Fort Collins, 1974.

23. Takao Kobayoshi and Stephen L. Sargent, "A Survey of Breakage Resistant Materials for Flatplate Solar Collector Covers," ISES Conference, Fort Collins, 1974.

24. S. W. Moore **et al.**, **op. cit.**

25. Adapted from tables by C. Clarkson and J. S. Herbert, **op. cit.**, and T. Kobayoshi and S. L. Sargent, **op. cit.**

26. Farrington Daniels, **Direct Use of the Sun's Energy** (New York: Ballantine Books, 1975), p. 58.

27. Adapted from tables by C. W. Clarkson and J. S. Herbert, **op. cit.**, and H. D. Young, **Fundamentals of Mechanics and Heat** (New York: McGraw-Hill, 1964), p. 494.

28. T. Kobayoshi and S. L. Sargent, **op. cit.**

29. (Note: Plexiglas and Lexon are product names for two of the basic sheet-plastics available—polymethyl methacrylate and polycarbonate. There are differences between the two types of plastic but the differences between the plastics are relatively small when compared with the differences between plastic and glass.)

30. C. W. Clarkson and J. S. Herbert, **op. cit.**

31. T. Kobayoshi and S. L. Sargent, **op. cit.**

32. F. Daniels, **op. cit.**, p. 58.

33. J. I. Yellot, **op. cit.**, p. 32.

34. F. Daniels, **op. cit.**

35. T. Kobayoshi and S. L. Sargent, **op. cit.**

36. John Keyes, **Harnessing the Sun** (Dobbs Ferry, New York: Morgan and Morgan, 1975), p. 46.

John Keyes uses a different definition for efficiency. He divides by the amount of solar energy transmitted through the coverplate instead of the amount incident on the covers. With this obviously smaller denominator, the "efficiencies" for his collector system are noticeably higher than normally understood.

37. Adapted from J. Yellot, **op. cit.**, p. 37.

38. Adapted from H. C. Hottel and B. B. Woertz, "The Performance of Flat-Plate Solar Heat Collectors," **Transactions of the ASME** (Vol. 64, 1942), pp. 99-100.

39. National Bureau of Standards, "Interim Performance Criteria for Solar Heating and Combined Heating/Cooling Systems and Dwellings" (January 1, 1975), p. 3.

C. D. Engebretson and N. G. Ashar, "Progress in Space Heating with Solar Energy," **ASME Paper No. 60 WA-88** (December, 1960), p. 3.

40. J. I. Yellot, **op. cit.**, p. 37.

41. S. W. Moore **et al.**, **op. cit.**, p. 24.

42. Adapted from D. J. Close, "Solar Air Heaters," **Solar Energy** (Vol. 7, No. 3, 1963), p. 123.

43. Adapted from H. Tabor, "Selective Surfaces for Solar Collectors," **ASHRAE, Low Temperature Engineering Applications of Solar Energy**)New York: ASHRAE, 1967), p. 46.

44. Figures 3.36 and 3.38 - 3.40 are from J. I. Yellot, **op. cit.**, p. 34.

45. Figure 3.37 from unpublished research supported by Londe-Parker-Michels, Inc., 1975.

46. C. D. Engebretson and N. G. Ashar, **op. cit.**, p. 6.

47. D. J. Close, **op. cit.**, pp. 117-124.

48. D. S. Ward and G. O. G. Lof, "Design and Construction of a Residential Solar Heating and Cooling System," **Solar Energy** (Vol. 17, No. 1, 1975).

49. **ASHRAE Handbook and Product Directory, 1978 Applications** (New York, American Society of Heating, Refrigerating and Air-Conditioning Engineers, Inc., 1978), p. 58.18.

   D. J. Close, **op. cit.**, pp. 117-124.

50. J. F. Kreider and F. Kreith, **Solar Heating and Cooling**, (Washington, D.C., Hemisphere Publishing Corporation, 1975), p. 142.

51. Adapted from table by H. D. Young, **Fundamentals of Mechanics and Heat** (New York: McGraw-Hill, 1964), p. 423.

52. S. W. Moore **et al.**, **op. cit.**, p. 10.

53. R. W. Bliss, Jr., "The Derivations of Several 'Plate-Efficiency Factors' Useful in the Design of Flat-Plate Solar Heat Collectors," **Solar Energy** (Vol. 3, No. 4, 1958), pp. 55-64.

54. A. G. Spanides, **Solar and Aeolian Energy** (New York: Plenum Press, 1961), p. 118. Published proceedings of International Seminar on Solar and Aeolian Energy Held at Sounion, Greece, September 4-15, 1961.

55. S. W. Moore **et al.**, **op. cit.**, pp. 3-4.

56. R. W. Bliss, Jr., **op. cit.**, p. 55-64.

57. S. W. Moore **et al.**, **op. cit.**, p. 13.

58. **Ibid.**, pp. 12-14.

59. Absolute temperature metric:
   °K = °C + 273 (degrees Kelvin)
   Absolute temperature English:
   °R = °F + 460 (degrees Rankine)

60. C. B. Eaton and H. A. Blum, "The Use of Moderate Vacuum Environments as a Means of Increasing the Collection Efficiencies and Operating Temperatures of Flat-Plate Solar Collectors," **Technical Program and Abstracts** (Fort Collins, Colorado, ISES Annual Meeting, August 20-23, 1974), p. 12, and personal notes.

   Jack Decker, "Demonstration of Solar Energy Utilization," Brochure of Solar Systems, Inc., Tyler, Texas.

61. NASA, "A Practical Solar Energy Heating and Cooling System," **Technical Brief B73-10156**, Marshall Space Flight Center (May 1973), p. 12.

62. **Ibid.**, pp. 27-28. Figures Adapted.

63. C. B. Eaton and H. A. Blum, **op. cit.**

64. Xerox of unpublished computer printout Section 203, Existing Construction 1973, U.S. Department of Commerce, Social and Economic Statistics Administration, Bureau of the Census.

65. Edmund Scientific Co., Catalog 751, p. 18.

66. D. Langer, "Characteristics of Cadmium Sulfide Photovoltaic Cells," **Solar and Aeolian Energy**, Proceedings of the International Seminar on Solar and Aeolian Energy, Sounion, Greece, Sept. 4-15, 1961 (New York: Plenum Press, 1964), pp. 234-235.

67. NASA, **Tech Brief B74-10090**, Langley Research Center (August, 1974).

68. Charles E. Backus, "Photovoltaic Program in the United States," Paper presented at Symposium "Solar Utilization Now," Jan. 13-19, 1975, at Arizona State University.

69. Karl Boer, "Part 1: The Solar One Conversion System," **Solar One: First Results** (Newark, Delaware: Institute of Energy Conversion, University of Delaware, 1974), pp. 5, 9.

70. NSF/NASA Solar Energy Panel, **Solar Energy as a National Resource**, U.S.G.P.O. Doc. 3800-00164, p. 56. Adapted from Figure 13.

71. Karl Boer, **op**. **cit**. Adapted from p. 9, 19.

    On a typical summer day, 1000 sq. ft. of solar cells with 6% conversion efficiency will produce approximately 40 kW/day. The average house is rated at 60.5 kW/day for this paper instead of 108.5 kW/day as implied by the Figures in the Solar One study.

72. **Ibid**., p. 8.

73. Charles A. Miller, "A Low-Cost Solar Cell Is Here!", **Mechanix Illustrated** (April, 1978), pp. 55-57.

CHAPTER 4

SOLAR ENERGY STORAGE

GENERAL

The basic intent of storage is to try to hold the heat energy made available by the sun until it is needed. The sun only makes its energy available for a limited number of hours during the day, depending on time of year and latitude (Figure 4.1). These limited hours of sunshine, on a reasonably clear day, will deliver energy to a collector similar to Curve 1 in Figure 4.2. The demand for heat in the dwelling is shown by Curve 2. The shaded difference between Curves 1 and 2 represents the amount of energy that can be diverted to storage to satisfy later demands for heat.

How long a system should hold its energy, how well it controls the release of it, and how much energy it should be able to hold are all design decisions that have to be made. And, just as with collectors, there are varying levels of technological response.

Most storage systems in use rely on the **specific heat** of the storage medium to provide thermal storage. There are, basically, two approaches to utilizing specific heat thermal storage systems. If the systems are required to hold and store the same

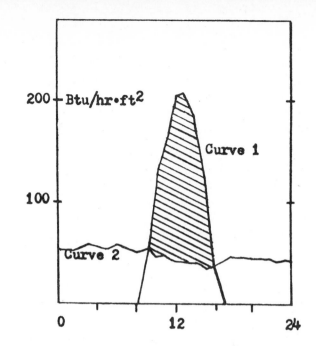

²**Figure 4.2**

Length of day in various northern latitudes (In hours and minutes on the 15th of each month)

Month	0°	10°	20°	30°	40°	50°	60°	70°	80°	90°
Jan.	12:07	11:35	11:02	10:24	9:37	8:30	6:38	0:00	0:00	0:00
Feb.	12:07	11:49	11:21	11:10	10:42	10:07	9:11	7:20	0:00	0:00
Mar.	12:07	12:04	12:00	11:57	11:53	11:48	11:41	11:28	10:52	0:00
Apr.	12:07	12:21	12:36	12:53	13:14	13:44	14:31	16:06	24:00	24:00
May	12:07	12:34	13:04	13:38	14:22	15:22	17:04	22:13	24:00	24:00
June	12:07	12:42	13:20	14:04	15:00	16:21	18:49	24:00	24:00	24:00
July	12:07	12:40	13:16	13:56	14:49	15:38	17:31	24:00	24:00	24:00
Aug.	12:07	12:28	12:50	13:16	13:48	14:33	15:46	18:26	24:00	24:00
Sept.	12:07	12:12	12:17	12:23	12:31	12:42	13:00	13:34	15:16	24:00
Oct.	12:07	11:55	11:42	11:28	11:10	10:47	10:11	9:03	5:10	0:00
Nov.	12:07	11:40	11:12	10:40	10:01	9:06	7:37	3:06	0:00	0:00
Dec.	12:07	11:32	10:56	10:14	9:20	8:05	5:54	0:00	0:00	0:00

¹**Figure 4.1**

quantities of usable energy, then most passive approaches could be characterized by their large storage mass, small temperature ranges and low operating temperatures. Active storage systems can generally be characterized by their small storage mass, large range of temperatures and significantly higher operating temperatures.

Passive storage temperature ranges hover around the space temperature that is desired. It is important that the storage system **be** part of the environment it is heating. Active storage temperatures begin around 32°C (90°F) and can easily reach 70°C (160°F). These high temperatures make it important to thermally isolate active storage components from the space it is heating. As systems, the higher temperatures of active storage schemes promote much greater rates of unwanted heat loss across the boundaries of the storage unit just because of the higher ΔT. Typically active storage systems are relatively small in volume; so, while they have large surface to volume ratios, the absolute amount of surface area is generally small relative to passive storage systems. Passive systems, however, have the advantage of operating at much lower temperature differences with respect to the outside environment, so the rates of heat loss per unit area are relatively low. In fact they would be only slightly more than the heat loss rate of the envelope itself—if it replaces the envelope. However, if the thermal mass of a passive storage system is disconnected from the envelope, there is, essentially, no heat loss to the environment.

In general, active storage systems operate at high ΔT across small amounts of area and passive storage systems operate across large areas at much lower ΔT—especially if consideration of the ΔT is limited to the marginal increases in wall temperature over a conventional wall in the same location: 5-7°C (9-13°F).

PASSIVE

The thick adobe walls of the arid southwest utilize their large thermal mass to even out the energy supply and demand curves. They hold heat for a short period of time, radiating towards both the interior and exterior environments. The only control with this design is in the thickness of the wall. Michel's design for Montmedy, France (Figure 3.13) uses the same thermal wall mass principle, but the design incorporates more control features. There is the glass cover that helps trap radiation that would have been lost to the exterior environment. The other three walls in the house have been designed to retard heat loss as much as possible, within economic limits. This thermal wall design incorporates vents which allow natural convection to distribute the heat and reduce the temperature difference at the skin; these vents can be closed to control heat distribution from this storage wall if need be—then the primary mechanism for heat delivery from storage is reduced to radiation to the interior environment. There are radiative losses from the glass wall to the exterior environment, but this extra "skin" helps retard such losses through the greenhouse effect. The glass will not transmit the longwave radiation from the wall, the glass must first absorb the energy before it can radiate it. The losses from the glass end up being significantly less than those incurred by an exposed absorbing wall.

Steve Baer's "drum wall" (Figure 3.16) is another passive storage response. His design however, goes one step further with control in that he has an insulating panel that covers the glass wall at night to reduce radiant heat loss from the glass.

Harold Hay's Atascadero residence (Figure 3.15) has moved the storage system to the roof of his structure and it utilizes insulating panels like the Baer design. The Hay house, however, has gone still further in that the design also has interior mass walls that are concrete block filled with grout or sand. These walls add substantially to the thermal mass of the dwelling and they function as a secondary storage system by slowly absorbing heat radiated down from the roof. These interior walls will then radiate their heat when the roof storage system is depleted. These interior, thermal walls add a lot more thermal inertia to the whole house; that is, extreme changes in the exterior environment will only slowly affect the temperature inside the dwelling. Another feature of the Hay system is that it also works well to store coolness in the summer.

The David Wright House (Figure 3.18) has perimeter structural storage walls as well as a storage floor. The storage walls represent a lot of area with a marginally small increase in ΔT compared to a lightweight insulating wall. The storage floor, except for its edges, can be considered internal mass that has relatively no heat loss to the outside environment.

The above storage system examples represent what can be done simply through good design. The **whole building** can be used to "passively" absorb and cope with the dynamic thermal impact of the natural environment.

The Hay system has provided its owner with 100 percent of the space heating and cooling requirements for the dwelling since it has been occupied. The particular micro-climate at Atascadero is relatively mild, but this design response has eliminated the need for furnace, air-conditioner, ducts, etc. Michel's system provides 75-80% of the energy necessary for heating, and has saved 70% on electric heating costs. This is in a relatively cold climate with only moderate levels of sunshine. The Baer's have a fairly cold heating season, too, and find that they are able to get most of their heat from the drum wall. For supplementary heat they utilize fireplaces and wood. The Wright House is in a cold region (6000+ DD per year) and it manages to get 90% of

its heating requirements through passive solar techniques.

It should be apparent that these passive solar buildings are performing at, or above the performance level expected of "reasonably" cost-effective active solar heating systems—60-80%.[3]

In terms of Passive Solar Storage, there is a much neglected subject: How much mass is enough? The general consensus is that there is no problem with excess mass: there cannot be too much mass. However, control of comfort (overheating) can be difficult and can reduce the effectiveness of the solar heating system if there is too little mass. From work done at Los Alamos Scientific Labs, it can be inferred that the **minimum** amount of storage for Trombe-type approaches should be at least 255 watts/°C per square meter of collector glazing (45 BTU/°F · Ft²). This assumes that all mass is directly irradiated by the sun and that there is only an insignificant mass throughout the rest of the house.[4] Again, this number is a minimum. No recommendation was alluded to in the Los Alamos work for a Direct Gain approach. However, the dynamics of controlling comfort in a sun-space should be facilitated by significantly increasing the mass and conductive surface area surrounding the sun-space: it is suggest that doubling the storage capacity may be warranted [510 watts/(°C · m²) or 90 BTU/(°F · Ft²)]. Until there is more definitive research, the maxim that "you can't have too much," should inform the judgement of designers while they use the above numbers as a lower, minimum bound on the design solution.

Because of the generally large differences in cost between active and passive heating systems, it is quite important that the designer first attempt to solve the problems passively: design so that the problem disappears.

ACTIVE

If the above systems perform so adequately the question has to be asked: why bother with all the expense and paraphernalia involved with installing and maintaining an "active" solar heating and/or cooling system? One basic reason would be the fact that the passive systems deliver their heat at relatively low temperatures; temperatures that are not adequate for domestic hot water or process heating loads. And, although solar cooling will probably not be cost-effective, absorbtion cooling and solar mechanical cooling work much better at higher temperatures. Therefore, active systems with relatively high storage temperatures are required. Other reasons probably involved are:

- familiarity — some passive approaches would demand changes in lifestyle and in concepts of what a house should be. These would require time to become culturally acceptable on a large scale;

- comfort — some people do not believe they could be comfortable with a range of temperatures, they feel they require the more precise control of a mechanical system with a narrow range theromostatically controlled;

- an American predisposition — solutions will be found in ever higher levels of technology, rather than in simple elegant design; and

- retrofit — some existing homes do not have unobstructed south facing walls, solar collection may be possible at the roof level or some other location remote from the building. These same physical limits could apply in new construction as well.

Before talking about actual active storage systems, a short outline of basic heat storage methods would put things in perspective—or a sort of performance hierarchy. There are two basic modes of thermal storage; one is dependent on the **specific heat** of a material; the other depends on its **heat of fusion**. The heat of fusion approach is only infrequently used. However, it may have large, passive implications in the future.

Specific heat, as a storage method, is a function of the sheer mass, or weight of material and its temperature. The amount of energy in storage at any one time is directly related to storage temperature (Figure 4.3). In order to calculate the amount of usable energy in the storage mass it is necessary to determine the lowest feasible operating temperature for storage. This is the limit where the temperature in the system has dropped too low for it to effectively supply space heat. Part of the results from the study of MIT's Solar IV House (1960) indicate that at around 30°C (85°F) it was more expensive to try to extract further heat from the active storage system than it was to turn on the auxiliary system to meet the

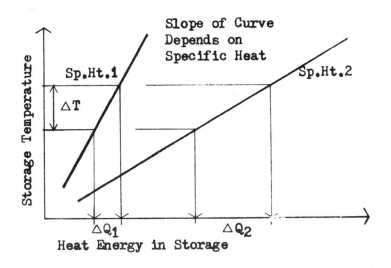

Figure 4.3

heating load for the dwelling.[5] With today's higher cost of electricity for pumps and fans, this minimum temperature may now be significantly higher, if the "auxiliary" fuel is not electricity as well. Any heat that is stored above this temperature cutoff is considered usable, so the amount of heat in storage becomes:

(specific heat) (weight of material) (ΔT)

In the Metric system:

$$\frac{\text{(calories) (grams)}}{\text{(gram °C)}} (\Delta C) = Q = \text{calories}$$

In the English system:

$$\frac{\text{(Btus) (pounds)}}{\text{(pounds °F)}} (\Delta F) = Q = \text{Btus}$$

NOTE: ΔT is the difference between storage temperature and 30°C (85°F)

Within this "specific heat" storage method there are two common types of storage media or material. One is water and the other is rocks. Now the "rocks" can be adobe or concrete, or sand in concrete blocks—the idea is that the storage media is a solid that warms up.

With an active collection and storage system, there is a discreet place called storage. This is a place where the heat collected, which is not immediately used for heating purposes, is held. It is generally within the dwelling; therefore, some of the heat lost from storage may help with heating the house. The fact that it is inside requires that some consideration be given to the actual placement of storage, since it will be a "hot zone" in the structure, probably very similar to a kitchen or laundry. It is also important to think about the structural requirements, since this storage unit will be delivering some pretty concentrated loads to whatever has to support it. Storage will also consume a relatively large volume, especially with the access space needed for all manner of pipe and duct connections that have to converge at the storage unit.

In order to compare the magnitudes of differences between rock bins and water tanks (the typical solutions), a quick, crude example of a 100 m² (1000 ft²) house will be discussed. The system is sized to provide 60-70% of the space heating requirements.

Although many articles in the past have suggested a rule of thumb of 1 sq. ft. of collector for every two feet of floor area to be heated—for a 60+% solar heated house, it should be obvious that the heat loss characteristics as well as the differences in climates for specific homes makes such a generalization virtually meaningless. Houses that are fairly well insulated should reduce this figure to as little as 1:5 or 6. The C.S.U. house, which was "typically" insulated, has a ratio of 1:4 with part of the heated space below grade. However, for this example, a collector area of 50 m² (538 ft²) will be used.

A rule of thumb for storage associated with a 60-70% solar house is that there should be 50-100 kilograms of water for each square meter of collector (10-20 pounds/ft² or 1.25-2.5 gals/ft²).[6]

Another rule of thumb with water storage systems is that a temperature swing of $\Delta C = 45°$ ($\Delta F = 81°$) is fairly reasonable. This sets the temperature range of storage at 30-75°C (85-166°F) and it will set the working limits on the thermal capacity of the storage tanks.[7]

Using 80 kg/m² (16.39 lbs/ft²) for the example, and knowing that the specific heat of water is 1.0 calories/gram°C (1.0 Btu/pound °F),[8] the size of the storage system can be determined. Figure 4.4 presents the results of the calculations.

	Metric	English
Collector Area	50 m²	538 ft²
Water/Collector Area	80 kilograms/m²	16.39 lbs/ft²
Total Water in Storage	4000 kg.	8816 lbs.
Weight/Unit Volume	1000 kg/m³	62.5 lbs/ft³
Least Storage Volume	4.0 m³	141 ft³
Specific Heat	1.0 cal/gm. °C	1.0 Btu/lb. °F
Temperature Range	$\Delta C° = 45°$	$\Delta F = 81°$
Heat Capacity	180,000 kcal (209 kW)	714,000 Btu

Figure 4.4

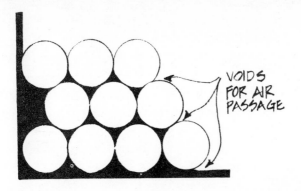

Figure 4.5

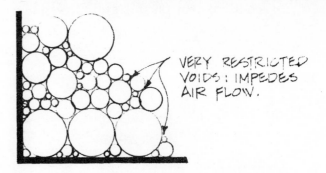

Figure 4.6

For rock bins, the procedure will be reversed in order to determine any likely rules of thumb, since there do not seem to be any readily available in the literature. The specific heat of rocks is about (0.2 Btu/pound °F).[9] In addition, the rocks should be carefully screened so that they are all the same size (Figure 4.5); because if they varied in size small rocks would fill in the gaps between the larger rocks (Figure 4.6) and this would severely cut down or completely cut off the flow of air among the rocks. Typically, the diameters of round gravel used ranged from 3.7cm (1.5 inches) to 7.5cm (3 inches), but they should all be the same diameter for any one application.[10] When packed in a bin, they leave approximately 30 percent of the volume void, so when calculating the mass of rock storage based on the weight/volume of solid rock—about 2400 kg/m³ (150 lb/ft³)—the weight required should be multiplied by 0.7 to yield the correct weight/volume of the screened gravel.

With the above information, the weight and volume of rocks necessary to store the same amount of heat as is in the water tank can be determined. The parameters for the rock storage system, meeting the same parameters as the water storage system, are shown in Figure 4.7. Rocks become a massive physical problem when compared to water storage. Rock storage has the disadvantages of weighing five

times as much as water and of requiring 2.6 times the volume to store the same amount of heat. These become real architectural constraints. If a bin were built with 2.2 meter (7.25 ft.) sides, this would be a structural load of over 4000 kg/m² (820 pounds/ft.²). This would seem to indicate that the storage unit should be placed somewhere where it could bear on footings of its own, perhaps to be structurally independent from the rest of the dwelling so that if it should settle at a different rate from the rest of the structure, it will not introduce undo stress on the rest of the foundation. Allowing some additional space for insulation, ducts, and access around the bin, which is 2.2 × 2.2 meters (7.25 ft. × 7.25 ft.), puts the space requirement in the same range as a FHA minimum 2.4 × 3.0 meters (8 ft. × 10 ft.) bedroom.

From these numbers, it is possible to backout a rule of thumb ratio for collector area to storage

	Metric	English
Heat Capacity	180,000 kcal (209 kW)	714,000 Btu
Specific Heat	0.2 cal/gm.°C	0.2 Btu/lb.°F
Temperature Range	ΔC = 45°	ΔF = 81°
Heat Stored/°ΔT	4000 cal/°C	8816 Btu/°F
Total Rocks in Storage	20,000 kg.	44,080 lb.
Weight/Unit Volume	1683 kg/m³ (30% void)	105 lb/ft³
Least Storage Volume	11.88 m³	420 ft³
Collector Area	50 m²	538 ft²
Rocks/Collector Area	400 kg/m²	82 lb/ft²

Figure 4.7

weight for rocks similar to the 50-100 kg/m² (10-20 pounds/ft.²) for water. For the rocks, these numbers are 300-485 kg/m² (65-100 pounds/ft.²). The example in Figure 4.7 came out at 400 kg/m² (82 lb/ft²).

Despite the above disadvantages, there are some real reasons to consider rocks over water as a storage medium. With rocks there is never any danger of freezing or leaking. There are no corrosion problems. The cost of air collectors, which normally would accompany a rock storage system, hold greater hopes for achieving low cost than do the water type collectors; so the entire system might be much less expensive—since collectors typically represent over 75 percent of a system's cost in a water collection scheme.[11]

Heat of fusion, or **latent heat**, as a storage mechanism, makes it possible to store much larger quantities of heat in equivalent weights or volumes of material than can be stored by using the specific heat, or sensible heat, mechanism. To use the heat of fusion for thermal storage it is necessary to find a material that melts—changes from a solid to a liquid—at some temperature within the expected storage swing temperature. Changing states provides a large, energy storage reservoir—at no increase in temperature—as the material soaks in the extra energy it needs to cross the energy threshold to become a liquid (Figure 4.8). The trick is to find compounds or mixes with high heats of fusion that melt at usable temperatures, that can cycle from solid-to-liquid-to-solid an indefinite number of times without their performance deteriorating, and that are readily available at low costs.

Salt hydrates are compounds that generally have high heats of fusion. Typically, they have latent heats around 55 cal/gm (100 Btu/lb.) and their specific heats are about 0.4 cal/gm°C while it is a solid and 0.7 cal/gm°C after it has changed into a liquid.[12]

The storing power of the heat of fusion method is best demonstrated by constrasting it to storage that

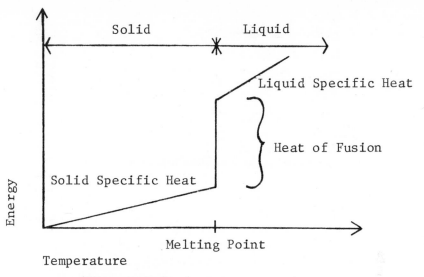

Figure 4.8

only uses the specific heat of the material. If the same 55 cal. were to be stored at 0.4 cal/gm°C, then, for the same weight, it would require a temperature rise of 137.5°C (250°F) before the same amount of energy was stored. To add this temperature rise to the minimum operating temperature of storage, places the temperature range required of storage beyond the possible delivery temperatures of flat plate collectors. Figure 4.9 illustrates this point and shows that the same weight of water would also require a temperature range beyond the effective delivery temperature of collectors.

This example illustrates the point that if weight or volume has to be limited in designing storage—for whatever reason—then it is important that the heat of fusion storage method be thoroughly investigated, since it results in tremendous weight and volume savings.

The following comparative example should show the relative advantages of rocks, water storage by specific heat and heat of fusion storage by salts.

If the problem were to try to store as much heat as possible at approximately 32°C (90°F), then sodium sulphate dehydrate ($Na_2SO_4 \cdot 10H_2O$) which melts

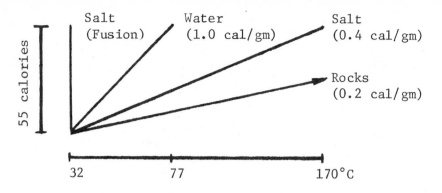

Figure 4.9

the amount of heat that the water would and 26 times the amount of heat in the same volume of rocks. By weight, the salt stores 6 times as much heat as water and 30 times as much heat as the rocks.

Figure 4.11 is a listing of the pertinent characteristics of some salts thought to be most suitable for heat of fusion storage. Note that some have relatively low melting points. These salts are used to withdraw heat from the environment at or below room temperature, in this way they provide a cooling effect and end up storing "coolness" removing heat. These could work well with efforts to achieve solar air conditioning during the summer months, both passively and actively.

These hydrated salts are chemical compounds which have a definite amount of water in their crystalline structures. They melt at precise temperatures and dissolve in this water of hydration. However, after melting they sometimes have trouble recrystallizing and releasing the energy that was stored when the crystalline structure "dissolved."

at 32.38°C (89.8°F) would be a likely choice. Figure 4.9 presents the characteristics of water, rocks, and the sodium salt along with comparative performance details. For a scenario, it will be assumed that the collector is able to deliver 38°C air to storage and that storage temperatures remain usable down to 27°C (80°F). This yields a ΔC of 11° ($\Delta F = 20°$) which straddles the 32.38°C (89.8°F) melting point of salt. The results in Figure 4.10 show that in comparing heat storage by volume the salt will store 8.9 times

		Weight/Volume	Specific Heat	Heat Stored by Temperature Range		Heat Stored by Weight	Heat Stored by Volume
METRIC				27 -- 32 -- 37°C			
	$Na_2SO_4 \cdot 10 H_2O$ Glauber's Salt	1460 kg/m³	0.42 cal/gm.°C (s) 0.79 cal/gm.°C (l)	2.1 55	3.95	61 cal/gm	89 kcal/m³
	Water	1000 kg/m³	1.0 cal/gm.°C	5.0	5.0	10 cal/gm	10.0 kcal/m³
	Rocks	1683 kg/m³	0.2 cal/gm.°C	1.0	1.0	2 cal/gm	3.4 kcal/m³
ENGLISH				80 -- 90 -- 100°F			
	$Na_2SO_4 \cdot 10 H_2O$ Glauber's Salt	91 lb/ft³	0.42 Btu/lb.°F (s) 0.79 Btu/lb.°F (l)	4.2 108	7.9	120 Btu/lb	10930 Btu/ft³
	Water	62.5 lb/ft³	1.0 Btu/lb.°F	10.0	10.0	20 Btu/lb	1250 Btu/ft³
	Rocks	105 lb/ft³	0.2 Btu/lb.°F	2.0	2.0	4 Btu/lb	420 Btu/ft³

Figure 4.10

	Melting Point °F	Heat of Fusion Btu/lb	Btu/ft³	Specific Heat Btu/lb.°F solid	liquid	Density lb/ft³
$NH_4Cl \cdot Na_2SO_4 \cdot 10H_2O$	52	70	6200	.32	.60	92
$NaCl \cdot NH_4Cl \cdot 2(Na_2SO_4) \cdot 20H_2O$	55	78	7200	.35	.65	92
$NaCl \cdot Na_2SO_4 \cdot 10H_2O$	65	80	7500	.42	.78	94
$Na_2SO_4 \cdot 10H_2O$	90	108	9830	.42	.79	91
$Na_2S_2O_3 \cdot 5H_2O$	119	90	9450	.35	.57	105
$NaC_2H_3O_2 \cdot 3H_2O$	136	114	9200	.47	.77	81

[13]**Figure 4.11**

This problem is called "super-cooling" and it can be overcome by introducing similar salts that melt at a higher temperature. This works because the added salt provides unmelted salt crystals as sites, or seeds, for the melted salt crystals to grow on; this encourages the salt to crystalize and release the stored heat.

There are other problems involved with hydrated salts. After a large number of melting cycles the chemicals can change and become chemicals which no longer have the properties required. Life-cycle testing of $Na_2SO_4 \cdot 10H_2O$ is currently underway at the Solar One research project. These tests should show whether salts can be expected to have a reasonably economic life span. Since the salts are stored in tubes (Figure 4.12), to increase the surface area for heat transfer, it will be necessary to find container materials—at reasonable cost—that can withstand the heat as well as the corrosive effect of the salts. Most metals succomb to the corrosion, and most plastics become brittle from the heat. However, improved plastics are being tested and evaluated; they are expected to solve this problem.

Most of the work done with heat of fusion has revolved around solving the problems associated with using hydrated salts. Other compounds, though, should be searched out and might not present such problems to begin with. Beeswax is an example of a relatively inert compound that has high heat of fusion (42.3 cal/gm); unfortunately it also has a relatively high melting point (62°C, 140°F) that might be above the critical portion of the collector delivery temperatures during the winter. However, beeswax still might be a very feasible storage medium, and there are probably other compounds around that

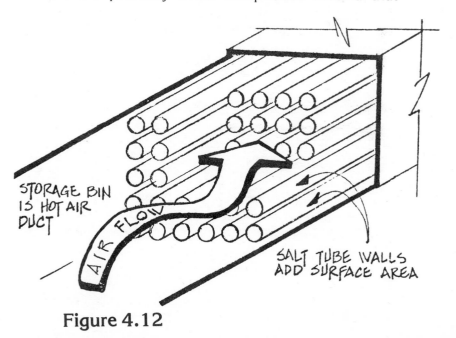

STORAGE BIN IS HOT AIR DUCT

AIR FLOW

SALT TUBE WALLS ADD SURFACE AREA

Figure 4.12

might be less trouble than the salts to utilize.

Salts do have a prominent place in these investigations because they are considered inexpensive and because different salts can be mixed, in varying proportions, to manipulate the melting point of the mixture—moving it up or down almost at will. This has a distinct advantage: almost all the heat can be stored at a temperature that the designer selects. In the above example, 90% of the heat was held at the precise melting point of the salt—32.38°C (89.8°F).

A particular mix associated with the idea of manipulating the melting point, by varying the different ratios of compounds, is the eutectic mixture. This mix is frequently referred to in the literature of heat storage. A eutectic mix is that specific ratio of compounds that produces the lowest melting point of all the possible variations on the mix (Figure 4.13).

In using the heat of fusion, it is most important to make sure that the melting point is sufficiently lower than the expected delivery temperature from the collector to insure efficient heat transfer; and that the melting point is sufficiently higher than the delivery temperature from storage to the heating load (space heat, hot water, or absorbtion refrigeration) to insure efficient heat transfer here as well.

The problem of thermal energy storage is rather straightforward. The technology, just as with collectors, is available. The choices of how to go about storage, however, can be complex; but they will basically center around economics. If space and/or weight have to be kept down for any reason, the designer would be advised to consider the heat of fusion method for storage. If weight or volume are not overriding concerns, the designer would be well advised to opt for rock storage with its simplicity and problem-free nature of operation. The real elegance, however, would come in utilizing some passive technique, whereby the structure of the building itself functions as collector, storage and distribution system all in one.

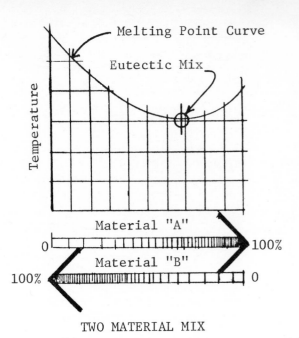

TWO MATERIAL MIX

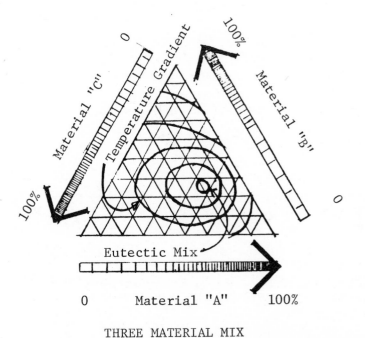

THREE MATERIAL MIX

Figure 4.13

In addition to being a buffer that helps to even out the curve of available energy, storage is also the buffer between the collector and the house; between energy acquisition and energy use. It is the interface for the distribution modes: from collector to storage and from storage to house.

There are two basic collection media: water and air. And there are two basic space heating media: water and air. Storage sits between them and must make the chosen combinations of the two compatible. The possible arrangements are illustrated in Figure 4.14 as a simple matrix.

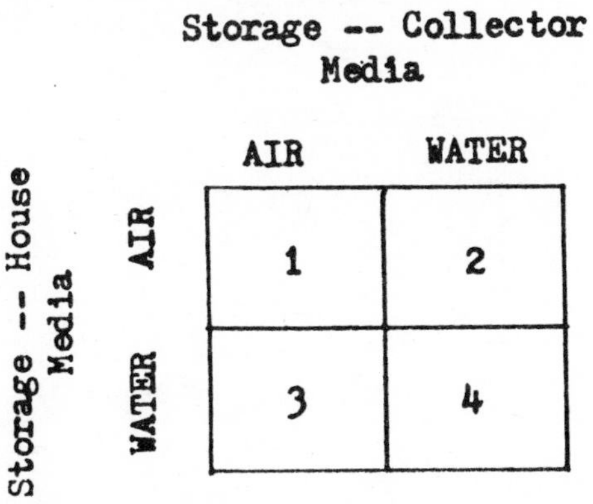

Storage -- Collector
Media

Figure 4.14

Element one is the matrix has storage as the interface for air collection and air distribution to the dwelling. There are two significant and different storage systems developed to date that interface these two systems. There is the system of rock storage developed by George Lof and employed in his Denver, Colorado residence, and there is a heat of fusion storage system developed by Maria Telkes which is in Solar One, the experimental house built by the Institute for Energy Conversion in Newark, Delaware.

In **the Lof Residence** 55.8m² (600 ft²) of collector of the type in Figure 3.39 are set at a 45° angle facing south on the flat roof of the one story, 186m²

(2000 ft.²) house. The air is forced down through the collectors and is then pulled to the basement. From here, it is either channeled straight into the house, much as in a typical forced air system with floor registers, or it is diverted to charge two 5.5 meter (18 ft.) towers filled with gravel 3.8-5.1 cm (1½-2 inches) in diameter. The towers are 0.9 meters (3 ft.) in diameter and they each hold 540 t (6 tons) of rocks.[14] Drawings of the disposition of the system are in Figure 4.15.

Denver has a little over 5500 degree days for a winter heating load, and this particular system has been able to provide 26 percent of the energy for a winter's space heat and hot water requirements. The air in the collectors has been able to reach temperatures in excess of 116°C (240°F), and the whole system has been in continuous operation since 1958.

Compared to what has been previously discussed, a serious question to ask is why does the system supply only 26 percent. The answer is probably a combination of several factors:

1. The area of collector seems rather small compared to the size of the house; it should be closer to 100m² (1000 ft.²).

2. The collector tilt is not quite optimum for winter heat collection; at 40°N latitude a tilt of 55-60 degrees would do better.

3. The size of the storage bins is much too small. The rules of thumb discussed earlier indicate that for this size house a storage of 40 tons, instead of 12, would be in order.

4. The house has a heat loss of approximately 25,000 Btu/DD, or 12.5 Btu/DD·ft². This is more than twice what it should be for good thermal design. However, when the house was built in the late 50's, added insulation was probably not cost effective at the time.

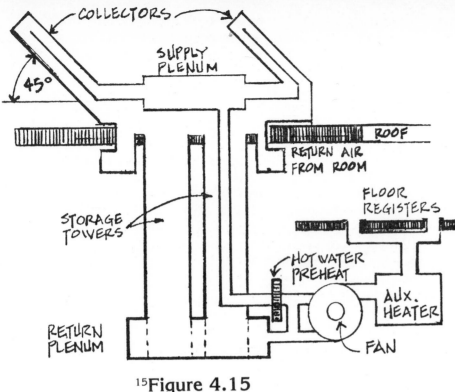

Figure 4.15

In other words, the solar system has been approximately sized to provide around 25% of the load. The whole system is scaled down compared to what has been discussed for 60-70% solar heating systems.

With the above notes as modifications, it would be reasonable to expect the system to supply around 70 percent of the winter heating needs. However, this says nothing about whether the economics to do so would be favorable or whether it is reasonable to provide the necessary 21.5m³ (760 ft.³) that would be necessary for the rock bins alone.

Solar One has twenty-four rooftop panels set at an angle of 45° and six vertical panels mounted in the south wall. These panels have an effective surface area of 70m² (750 ft.²) and is a little more than one half of the 120m² (1300 ft.²) living area; there is a full basement as well. The collectors are of the type in Figure 3.36. A forced air system cools the finned

plates, rising from the eaves to the ridge. It is then pumped through a heat storage bin filled with an hydrated salt: sodium thiosulphate pentahydrate, $Na_2S_2O_3 + 5H_2O$. The air from the collectors has to pass through storage to reach the forced air distribution networks to the house, the air cannot be diverted directly to the house.

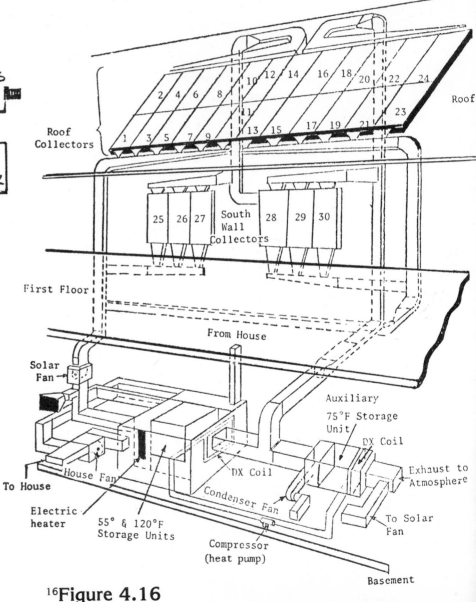

[16]Figure 4.16

There are 3.15 t (7000 pounds, 3.5 tons) of sodium thiosulphate with a heat of fusion of 47.9 cal/gram (90 Btu/pound) at 49°C (120°F). The salt is contained in 294 sealed pans, each 53cm (21 inches) square and 2.54cm (1 inch) thick. These pans are stacked in a volume of 2.32m³ (83 ft.³) so that 32 percent of the volume is void for air circulation. Since air requires large surface areas for effective heat transfer, there is 180m² (1940 ft.²) of surface area in the bin.

In addition, there are two other heat storage bins; one provides "cool" storage at 12°C (55°F), and the second provides storage at 24°C (75°F). It is coupled to a heat pump to boost the heat storage capacity of the 49°C (120°F) primary heat storage bin. Figure 4.16 is a schematic of the solar heating system.

This system has been operating in Newark, Delaware at 39.5°N latitude since July, 1973. The first winter ('73-'74), solar energy supplied 67 percent of the heating needs. The annual degree days for the area is about 4950 DD. Some modifications have been made and predictions are that the system will carry 80 percent of the heating requirements from now on.

The thermal storage capacity of the salts is large when compared to the Lof scaled down rock storage system. However, Lof's is a proven system; it has been in operation since 1958. The workability of the salt storage method is yet to be proven in either expected life span or economic feasibility.[17]

Element two in the matrix (Figure 4.14) is a water collection system with an air circulation system to the house. The water collection system virtually demands a water storage system. The forced air method of distribution to the house requires some large heat transfer surface, fed by the storage tank, with appropriate ducts and fans. There are three major examples of this water-air combination.

1. MIT Solar IV.
2. Thomason's House
3. Colorado State University (CSU) Experimental Solar House

Each system has nuances that are worth discussing.

The MIT Solar House IV is a two-story design with 135m² (1450 ft.²) of living area. It has 59.5m² (640 ft.²) of collector at a tilt of 60°. The collectors are copper tubes mechanically bonded to aluminum channels (Figure 4.17). The storage tank has a capacity of 5675 liters (1500 gal.) and it is backed up by a 378 liter (100 gal.) oil fired boiler.

The house was built in 1959 and was operated during the worst six months of the heating season for 59-60. The solar heating system carried 46 percent of the heating load. It is logical to expect that the system would produce a higher overall efficiency if it were operated during the full heating season. This six month period (October - March) was the period most deficient in solar radiation and it experienced more severe weather conditions than normal—5798 degree days compared to a normal 4895 DD for the same period.

Some of the important features of this system are:

It used copper pipe throughout, thereby avoiding corrosion problems.

It used city water without further treatment. The collectors were self-draining when the temperature dropped below storage temperature +3.5°F.

It was an open cycle system; the water in storage actually circulated through the collectors.

It had an oversized heat exchanger, compared to conventional heat exchangers in a forced air system, because the temperatures of the solar storage system were lower than conventional boiler temperatures and additional surface area on the exchanger was required for efficient heat transfer.[19]

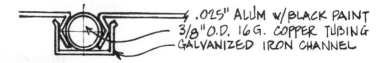

.025" ALUM w/BLACK PAINT
3/8" O.D. 16G. COPPER TUBING
GALVANIZED IRON CHANNEL

[18]**Figure 4.17**

The Thomason House. Harry Thomason is an entrepreneur, solar system developer who began devising solar heating systems in 1959. His systems are remarkable for their simplicity, but there is a lot of overkill in his approach to collection and storage. However, since there has been inadequate instrumentation to evaluate the performance of his systems and to correlate them with weather phenomena, his systems are viewed with some skepticism.

His first design was a 140m² (1500 ft.²) house of which 84m² (900 ft.²) was heated. The collector was 78m² (840 ft.²): almost a one to one ratio with the heated portion of the house. The collector design was rather inexpensive (about $1.00/sq.ft. in 1959) and is shown in Figure 4.18. The water transfer medium was distributed by copper piping to the top of the collectors and then it trickled down through open corrugated aluminum troughs to an insulated gutter. It was returned by gravity to a 1600-gallon steel tank through a filter that removed debris picked up from the gutter. The tank was surrounded by 45 t (50 tons) of fist sized rocks. This combined storage system is 2.5 times the size suggested by the rules of thumb (2 gallons/ft.² collector = 1680 gallons or 82 pounds of rock/ft.² = 34.4 tons of rock) and it oc-cupies more than 18.6m² (200 ft.²)—without considering the ducts and fans. Collector efficiencies were about 45 percent—the same as the MIT collector which operated in a much more severe climate (5800 DD for six months in Massachusetts compared to 4225 for the entire year near Washington, D.C.)

Some important features to note about this system are:

Heat is conducted from the water tank to the rocks which are used as a heat exchanger. The whole storage volume (1600 ft.³) is a heat exchanging plenum for air distributed to the house. There is lots of surface area. The operating temperature of the storage/exchanger was not reported.

There has been no report of corrosion of the aluminum absorber so the paint must protect the sheets from ion exchange with water from the copper pipes.

The system operates using city water with no additives. The collectors, of course, drain when the pumps cut off, since it is an open trickling collector. The collector heat transfer fluid (water) operates in an open-cycle circulation system with the storage water volume.

There is some gain in heat collection efficiency by virtue of the fact that the water tank rises to a higher temperature than the rock bin during the day. Overnight, the conduction of heat to the rocks levels out the temperature between the rocks and the water tank. With the water tank at a lower temperature in the morning, it can more efficiently absorb the energy from the collectors than it could if it had remained as hot as it had been the preceding evening. This allows the system to take advantage of more of the available energy than the MIT system, since collection can begin at lower temperatures.

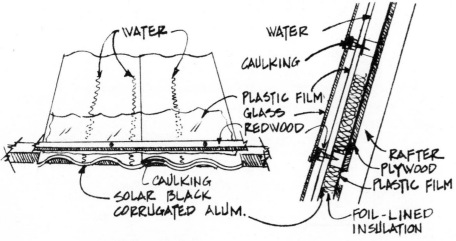

[20]**Figure 4.18**

Even with twice the suggested amount of collectors and 2.5 times the suggested storage capacity, relative to a solar installation capable of carrying 70 percent of the heating load, the Thomason system was still unable to carry 100% of the heating load. There are not any accurate statements of how much of the load was supplied by solar energy, but the first winter required only $4.65 of heating oil with solar heat compared to an estimated $100-$125 without it. This suggests that the system carried up to 95 percent of the load with an estimated five days of storage capacity. This also shows that five day storage is still not enough to carry the worst weather conditions.

Since only the most severe weather would be likely to completely exhaust storage, a full-scale, conventional heating system would still be required to back up the solar installation—even if it is for just 5% of a winter's heat. So, even with a 95% solar heating system, there will be no reduction in the cost of the auxiliary system. What is needed is an investigation to find an economic optimum. If it follows the pattern of recent investigations, the optimum will be around a 70% solar heating system.

During summer nights, water is trickled down the north side of the roof. This cools the water through night sky radiation and evaporation. The water returns to storage and cools the rocks by absorbing the heat away from them. The house is cooled and dehumidified by first blowing air through the cooled rocks and then circulating it throughout the house.

The roof is in two almost equal sections of tilt, one at 45°; the other at 60°; Figure 4.19. The reasons for this are not apparent.

The house is oriented 10° west of south to take more advantage of the afternoon sun; with generally higher ambient afternoon air temperatures, the collector is supposed to

operate more efficiently because there is less heat loss. This is an interesting and somewhat logical bit of design information, but there is no test data that would prove that this is the case; or, that if it is, it makes a significant difference.

There have been two subsequent Thomason solar houses of note, but the first is the basic design for both of them. In general, he has moved toward collector tilts of 60°. He has slightly reduced the ratio of collector area to heated house area somewhat, but he has extended his solar capture capability by placing reflective surfaces so that additional solar energy is directed to the collectors (Figure 3.16). In the third house, he has kept his water storage at about 1600 gallons, but he has reduced the rock fill by one half to 25 tons. However, he is using a concrete block water tank with a 2500 gallon capacity (3.75' × 3.75' × 25'); and his whole storage volume occupies more space—28m² (300 ft.²)—and volume

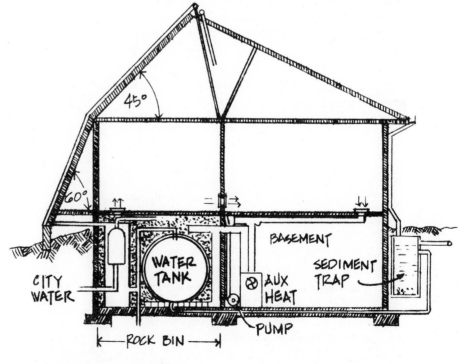

[22]Figure 4.19

—60m³ (2100 ft.³)—than the first design. This third design is supposed to be 75-85 percent solar heated.[21]

The Colorado State University (CSU) house in Fort Collins, Colorado is an experimental dwelling built under the research directorship of George Lof. The decision to use a liquid medium for solar collection and water for thermal storage was dictated by the fact that the house was to have a solar driven air conditioning system of the absorbtion-refrigeration variety. This system was judged to be most compatible with water storage. Water storage was also chosen for high heat capacity and low cost. The house is one story and has 140m² (1500 ft.²) living area with a full walkout basement for a total heated area of 280m² (3000 ft.²). The house is reportedly "normally" insulated, and it has a heat loss of 17,6000 Btu/Degree Day or 5.86 Btu/(Ft.² · DD). Heat loss for the winter design conditions of −23°C (−10°F) is 55,000 Btu/Hour. Heat loss at winter design conditions is primarily significant for sizing the load carrying capability of the auxiliary heating system.

The system has 16 roll-bonded aluminum panels (Figure 3.45) at a 45° tilt for a total collector area of 71.3m² (768 ft.²). This heats a 4275 liter (1100 gallon) storage tank with a volume 4m³ (143 ft.³) and a weight of 4.3 t (4.8 tons). Copper piping circulates 28 gallons of an antifreeze (ethylene glycol) and water solution between the aluminum collectors and two heat exchangers in the storage tank. The solar system in the C.S.U. solar house was designed to handle three fourths of the heating and cooling load of a typical 280m² (3000 ft.²) house. It has a standard gas fired boiler that has been sized, in the normal fashion, to serve the maximum heating or cooling demands of the building. The boiler is maintained at 90°C (190°F). Since this is a research project to gather data, a lot of alternate operating modes have been designed into the system. So, there are a lot of features that will be noted; but not all of them will be applicable to any one solution.

The collector is tilted to a 45° angle from horizontal. This represents a compromise between the optimum tilt for solar heating in the winter and the optimum tilt for solar cooling during the summer.

The circulating liquid is isolated from the 1100 gallon storage in a closed-loop system (as opposed to the two previous open systems). Advantages to the closed system are:

- This loop can be pressurized to raise the boiling point of the fluid so that higher collection temperatures can be achieved during the summer—to better meet the requirements of the absorption refrigeration unit.
- It is easier to accomplish the necessary filtration and de-ionization to prevent galvanic corrosion of the aluminum collectors (28 gallons versus 1100 gallons).
- This system requires much less antifreeze and chemical inhibitors to adjust the circulating fluid to the proper chemical concentrations.

If power ever fails or the pumps stop operating for some reason, a surge tank of 30 gallons is on the outlet side of the collectors in case the circulating fluid has to boil out of the collectors (summer stagnation temperatures can reach 204°C (400°F).

Because the circulating fluid is designed not to freeze, the collectors do not drain themselves. This might provide a path for heat conduction to the collectors during the night and could be a source of considerable heat loss. Whether the loss is actually significant or not has to be determined.

In order to prevent galvanic corrosion, it is

necessary that dissimilar metals be electrically insulated from each other in water systems. For example, flexible rubber hoses connect the copper piping to the aluminum manifolds. There is also a roll of aluminum screening in the rubber hoses that will sacrifice itself to any corrosive metal ions before they get to the aluminum collector.

If storage were operated as an open loop system with the circulating fluid, the absence of the heat exchangers, which are needed in a closed system, would boost performance to slightly higher efficiencies. This would result in approximately one percent higher solar contribution over the entire year.

Solar heating coils in air ducts are **followed** by the auxiliary heating coils. This dual coil system is slightly more expensive, but it allows use of stored solar heat even when it is not adequate to match house demand. This solar preheat coil ensures more use of solar and less use of auxiliary heat.

Total cost of solar installation over and above conventional cost was $6500, of which $5000 was for the collectors and their installation.

Air delivery to house is most compatible when there is air conditioning required as well as heating.[23]

There does not seem to be any available examples for **Element three** of the matrix (Figure 4.14). It seems that there are no air collection systems which operate with water as the distribution medium to the dwelling. Even further, there is no available example of an air collection and distribution system with water storage. However, it might well be an important system to investigate. It seems that there are a number of advantages that would result from such a system:

Air collectors are significantly less expensive than water collectors. And there is no problem with corrosion or freezing.

Water is a significantly better storage material than rocks.

This seems to be an ideal system; it has, to a large extent, the most advantageous features from the other systems without their major complications. The only difficulty that is foreseen is the development of an efficient way to transfer heat from the hot air to the storage water. Perhaps all that is needed is a system of coils in ducts similar to that in the CSU or the MIT IV houses, but operated in reverse so that they pump heat into storage. Perhaps fins around a tank, or a maze of heat pipes in a duct. Any one or all of these might be an answer to providing enough surface area, since the amount of surface area is critical to the effectiveness of heat transfer.

Although storage in this case is meant to be an interface between air collection and water distribution to the house, there seems to be no reason why this storage system could not also operate with an air distribution system as well.

A system that employs air collection and water storage should be tested to determine whether it would work well, since there do seem to be a lot of advantages associated with it.

Element four in the matrix matches a water collection system with a water distribution system. The most compatible storage is obviously water. There is one example of this storage interface. It is **MIT Solar House III**, the 1948 remodeling of the MIT Solar House II. The house was built in order to test the accuracy of theoretical solar heating models. In order to do this, the system was purposely "underdesigned." It had 37m² (400 ft.²) of collector tilted at 75° (latitude plus 15°) to optimize collection of winter sun. It had 4540 liters (1200 gallons) of water storage. The system was designed to provide 75 percent of the heat for the winter through a system of

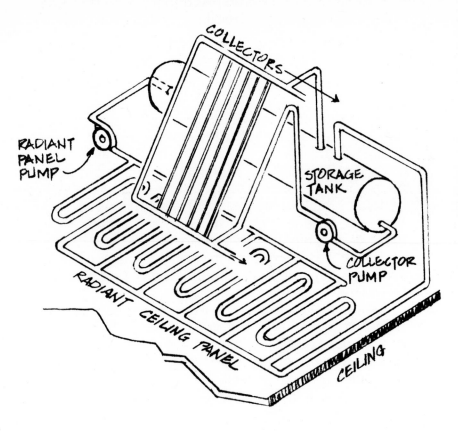

COLLECTORS

RADIANT PANEL PUMP

STORAGE TANK

COLLECTOR PUMP

RADIANT CEILING PANEL

CEILING

[24]**Figure 4.20**

hot water radiant panels in the ceiling of the 56m² (600 ft.²) living unit. The system performed very closely to the predicted levels and validated the theoretical assumptions and methods.

Figure 4.20 is a schematic of the house and its system. Because the system was part of a remodeling effort (rather than new construction) storage had to be placed in the attic. This meant that the tank had to be narrow in diameter and quite long to distribute the load over as many ceiling joists as possible. In addition, heat lost from the tank only served to warm the attic; it was of no help to the heat load of the dwelling.

In "Heating by Sunpower: A Progress Report", Professor A. L. Hesselscherdt, Jr., a member of the MIT solar research team concluded that, although the house in question (MIT III) was in a location where climatic conditions of a heavy heating load and poor atmospheric conditions did not favor solar heating, solar energy could be successfully used for space heating. He also noted that the design of an energy transport system is extremely critical and requires expert attention. This would be good advice to all those that would haphazardly approach solar implementation schemes. The heat transfer fluid must be distributed uniformly over the absorbing surface.

It is interesting to note that MIT Solar III was purposefully undersized to yield only 75 percent of the heating load. Since this report in 1948, a number of economic studies have suggested that to try to store enough heat to satisfy 100 percent of a winter's heating requirements is too costly. It has been determined that an essential requirement for a system is that it be sized so that it accumulates enough heat in the daytime for use at night. Depending on the climate, storage for overnight heating only will produce seasonal heating averages of close to 70 percent. If it is determined that storage is needed for longer periods of time, then the collector and storage may need to be expanded exponentially. Therefore, the 75% cutoff that was set in 1948 for the MIT III house fits well with recent economic optimization studies.[25]

It should be pointed out, however, that these economic studies were done several years ago, and the cost of fuel has risen substantially. Perhaps new studies should be undertaken to establish whether or not this solar optimum has shifted, especially since the only foreseeable prospect is that the cost of fuel will continue to rise.

Although active solar heating systems have generally been only marginally cost effective over the projected twenty year life of a system, the ultimate scarcity of conventional fuels and their price may well make the "manufacturing" of solar energy through active devices quite cost effective.

At the present time, utilizing active solar systems to produce hot water for process loads is quite cost effective—because of the year-long collection of energy.

NOTES

1. G. T. Trewartha, **An Introduction to Climate** (New York: McGraw-Hill Book Company, 1968), p. 13.

2. G. O. G. Lof and R. A. Tybout, "House Heating with Solar Energy, **Solar Energy** (Vol. 14, 1973), p. 259.

3. C. D. Engebretson and N. G. Ashar, "Progress in Space Heating with Solar Energy," ASME paper N. 60-WA-88 (December, 1960), p. 2.

   This document suggests that 75% of space heating is practical in the New England climate.

4. J. D. Balcomb and R. D. McFarland, "A Simple Empirical Method for Estimating the Performance of a Passive Solar Heated Building of the Thermal Storage Wall Type," presented at the 2nd National Passive Solar Conference, Philadelphia, PA, March 16-18, 1978.

5. C. D. Engebretson, **op. cit.**, p. 4.

6. **Ibid.**, p. 2.

   The utilization of 1500 gals/608 ft² of collector suggests that approximately 20 pounds per square foot of collector is optimum for their design and climate.

   G. O. G. Lof and R. A. Tybout, **op. cit.**, p. 276.

   This article suggests that 10-15 pounds of water (or its thermal equivalent) is the required heat storage capacity for minimum solar heating costs in practical situations—47-75% solar depending on climate.

7. G. L. Moore, "Sizing of Solar Energy Storage Systems Using Local Weather Records," 1975, p. 4.

8. M. Telkes, "Solar House Heating—A Problem of Heat Storage," **Heating and Ventilating** (May, 1947), p. 71.

9. **Ibid.**, p. 71.

10. W. A. Shurcliff, "Solar Heated Buildings: A Brief Survey," 5th Edition, (August 17, 1974) p. 11.

    Lof uses gravel of 1-1½" diameter p. 36. Thomason uses "fist sized stones" or roughly 3" diameter stones. The range of gravel sizes can vary somewhat in any one application, but this range should be kept as small as is practically possible.

11. G. O. G. Lof, etc., "Design and Construction of a Residential Solar Heating and Cooling System," NTIS Doc. No. PB237042 (August, 1974), p. 49.

    According to this study the collectors cost $5106.00 while storage cost $1400.00. This means the collectors were 78% of the cost of the system.

12. P. I. Altman, "Conservation and Better Utilization of Electric Power by Means of Thermal Energy Storage and Solar Heating," Interim NSF Report, NTIS Doc. No. PB210359 (Oct. 1971).

13. **Ibid**.

14. B. Anderson, "Solar Energy and Shelter Design," Master of Architecture Thesis MIT, (January, 1973) pp. 133-134.

15. P. J. Steadman, **Energy, Environment, and Building** (New York: Cambridge University Press, 1975), p. 127. Adapted.

16. T. M. Kuzay, **et. al.**, "Part 2: The House and Its Thermal System," **Solar One: First Results** (Newark, Delaware: Institute of Energy Conversion, University of Delaware, 1974), Figure 5, p. 24.

17. **Ibid**., pp. 1-18.

18. Figure from C. D. Engbretson and N. G. Ashar, **op. cit**., p. 3.

19. **Ibid**., data used.

20. Figure from Bruce Anderson, **op. cit**., p. 140.

21. Data on Thomason houses from Harry Thomason, "Solar Houses and Solar House Models" as reproduced in B. Anderson's Solar Energy and Shelter Design," **op. cit**., pp. 139-146.

22. **Ibid**., Figure adapted.

23. Information on the C. S. U. House from Lof, "Design and Construction of a Residential Solar Heating and Cooling System," NTIS Doc. No. PB7042 (August, 1974).

24. B. Anderson, **op. cit**., p. 16. Figure adapted.

25. **Ibid**., pp. 114-119. MIT III House data.

CHAPTER 5

SIZING

GENERAL

The purpose behind this book is to develop an information base for, and an approach to, using solar energy to heat a home. The goal is to collect and store heat. The amount of heat that needs to be stored to satisfy the criteria established by the designer will be the factor that determines everything else. Once this decision is made, the sizing and piecing together of a system is a fairly straight forward matter.

However, the decision of how much storage to carry is a matter of judgement. It will have to be an individual optimum based on the hard realities of climatic conditions and economic limitations. The basic question is: How much solar energy is actually used for every dollar spent on the system (BTU/$)? The following discussion will attempt to develop the conceptual framework from which the reader can make an informed judgement concerning the system he would require.

STORAGE DYNAMICS

There might be a tendency to ignore the importance of storage, because it is generally the least expensive component of the solar system, and because the major thrust of research and articles written in the field deal with the problems of collection. With this tendency, it might be easy to assume that the more storage available, the better. This is generally true for a Passive Solar System; but storage needs to be very carefully sized in an Active Solar System.

For example, a particular collector subsystem of fixed area on a given day, might be able to deliver a quantity of heat, Q, to an active storage system, at an average temperature of T_c. Figures 5.1 A, B, C graphically show the performance of an "optimum" sized storage, a storage unit twice the size of the optimum unit and a storage unit one-half the size of op-

timum. For comparison sake, it will be assumed that the temperature of all storage units start out the same and that the working temperature of storage has to be higher than a minimum temperature, T_h, that represents the needed temperature differential to provide efficient transfer of heat to the house.

In Figure 5.1 A, the optimum nature of storage can been seen in the fact that fully 75 percent of the quantity of heat supplied from the collectors is within a useable temperature range. Only 25 percent went toward charging the storage unit up to the minimum temperature. No heat had to be rejected in what is called "heat overflow."

In Figure 5.1 B, the fact that there is twice the storage material that has to be heated means that 50 percent of the heat delivered from the collector is used just to charge the unit to the minimum working

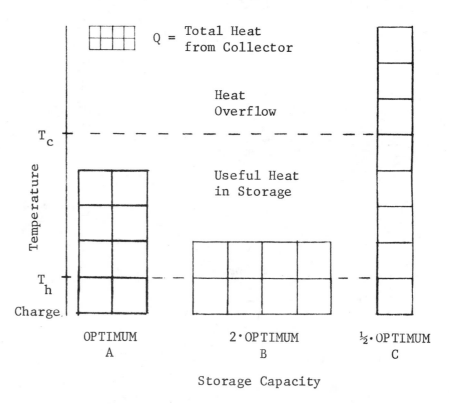

Figure 5.1

113

temperature. Because of this, only half the available heat from the collectors now lies in a usable temperature range. This means that under collector and weather conditions that supplied this quantity of heat to storage, the system with twice the storage of the "optimum" unit, can only supply 66 percent as much heat to the house load as the optimum can.

Figure 5.1 C shows the performance of a unit one-half the size of optimum storage. Since it's half sized, only 12.5 percent is required to charge the storage to a minimum working temperature. However, the storage unit will relatively quickly reach the delivery temperature of the heat coming from the collectors. At this point, there is no temperature difference; and, therefore, no transfer of heat to storage. In this case 37.5 percent of the heat supplied just passes through storage as heat overflow, because there is not enough storage capacity to absorb it.

In addition to the effect sizing has on storing usable energy, there are other nuances in performance that should be understood. First of all, if the storage unit was in operation prior to the start of this cycle, it was probably already charged to the minimum operating temperature; and, it would probably not have dropped too much below this temperature, because it would have stopped operating when it got to its minimum operating temperature. If this is the case, then the storage capacity of the optimum, and twice-optimum, shifts upward (Figure 5.2A, 5.2B) and both can store 100 percent of the heat, Q, delivered in a usable temperature range. Although they store all the available heat, Figure 5.2 A is still considered "optimum" because the heat it has is stored at higher temperatures and it offers more efficient transfer of heat to the house when it is called for.

If the outlet temperature from the collector had been held constant, then the example in 5.2 A would have required more pump or fan power to deliver energy into storage. This is a function of the rate of heat transfer. If the ΔT between collector and storage is small, more mass at T_c has to be circulated to keep a reasonable thermal pressure across the heat exchanger. Conversely, 5.2 B is sending a colder heat transfer fluid to the collector. Pump or fan power would be reduced because the fluid requires more time in the collector to rise to T_c.

If the collector outlet temperature is designed to be a set ΔT above the slowly rising storage temperatures, then case 5.2 B will use more pump or fan power. It will also actually collect more energy due to lower collector operating temperatures and the resulting higher efficiencies. Whether the extra energy collected is worth the energy used to collect it, is a function of the system components and design.

On the other side of the equation, a higher temperature in storage means that less energy is required for pumps or fans to extract this heat. And, if

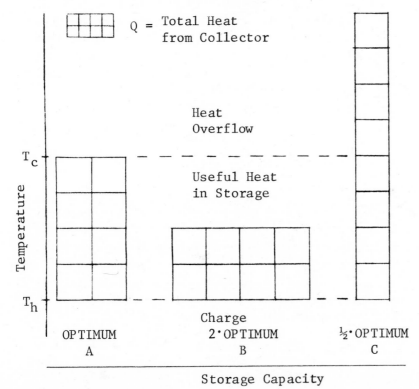

Q = Total Heat from Collector

Heat Overflow

Useful Heat in Storage

T_c

Temperature

T_h

Charge

OPTIMUM A

2·OPTIMUM B

½·OPTIMUM C

Storage Capacity

Figure 5.2

the temperature in storage is sufficiently high, it might be feasible to provide the domestic hot water requirements as well.

Heat loss from storage, while it waits, also needs to be considered in the system's overall efficiency. The storage unit with a higher ΔT would have a smaller, more economical to insulate surface area. Such a unit could easily have less overall heat loss than a unit with a lower ΔT, considerably more surface area and, consequently, relatively higher initial cost.

If the half-optimum system (5.2 C) starts storing at

the minimum temperature, it merely shifts the energy that would be lost from charging into the overflow category. Once it reaches the upper limit of the range, it cannot store any additional energy. The upper limit will be slightly lower than the expected collector operating temperature.

This points up the fact that not only is it necessary to decide how much storage is required; but also the temperature range over which it should be stored. It should be pointed out that if the heat of fusion technique is used, all the heat that is required can be stored at a precise temperature that is fixed by the designer, if a mixture can be devised or found that melts at this temperature.

Fixing the temperature range that a system needs to operate over requires an understanding of the dynamics of the system to make a reasonably good judgement on storage requirements.

Figure 5.3 is adapted from a study done by G. O. G. Lof and R. A. Tybout, "Cost of House Heating With Solar Energy". The curves show how different factors affect storage. Curve "I" represents the

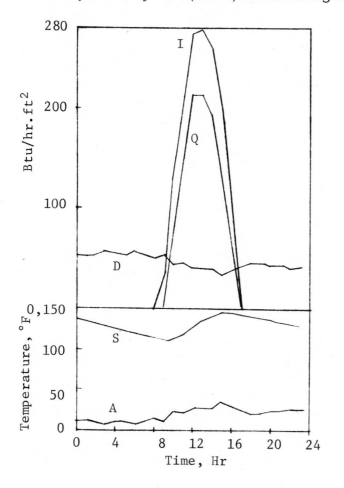

Figure 5.3

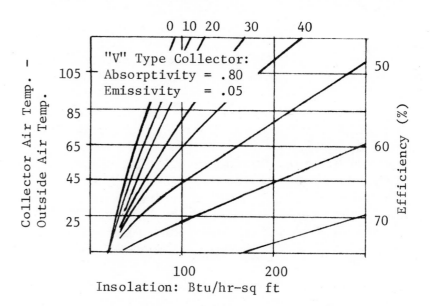

[5]**Figure 5.4**

115

number of Btu/hr that fall on a tilted collector surface.[1] Curve "Q" is the actual heat output available from the collectors. Curve "D" is the demand heat load for the residence and it is a mirror image of the outside air temperature, Curve "A", on which it is dependent. Curve "S" is the temperature in storage. Since the quantity of heat in storage is proportional to the storage temperature, Curve "S" also represents the amount of heat in storage.

From 0:0 hours to about 9:30 hours storage shows a continual decline, even though the sun has been shining on the collectors since about 7:30 hours. This can be explained by two reasons. The first is that, at its lowest point, storage temperature is about 49°C (120°F); and, according to the figures for a "best" air collector design[2] discussed by D. J. Close (Figure 5.4), an insolation level of approximately .339 cal/cm²·min (75 Btu/hr. sq. ft.) is required to produce a temperature in the collector of 61°ΔC (110°ΔF) above the outside air temperature. 61°ΔC (110°ΔF) was chosen because air temperature is at approximately −12°C (10°F) and storage is at 49°C (120°F). The insolation level has to be higher than .339 cal/cm²min (75 Btu/ft.²hr.) for the collector to be hotter than storage so that heat can be used for space heating or transferred into storage. This occurs shortly after 8:30, but storage still does not begin to charge. The second reason is the fact that until the heat output from the collectors is greater than the heat demand, storage continues to discharge heat—to make up the difference. At approximately 9:30 the insolation curve crosses the demand curve. From this point on, input from the collectors is greater than demand, and the collector temperature is higher than the storage temperature. At 9:30 Curve "S" begins to rise, indicating that storage is being charged.

Although the temperature of the collector was probably high enough at 8:30, storage did not charge until 9:30 when there is an excess of heat above demand. If demand had been lower, then the storage probably could have begun charging at 8:30, if storage had dropped to 49°C (120°F) by then.

The storage curve in Figure 5.3 shows that a maximum temperature of 66°C (150°F) is reached at about 1500 hours (3:00 p.m.) and the storage temperature levels off until 1600 hours when the energy supply curve from the collectors, "Q", drops below the rising energy demand curve, "D". Storage begins the discharge portion of its cycle at this point. The reason storage does not continue charging can be found, again, in Figure 5.3. The outside air temperature has risen to about 0°C (32°F) and storage has reached 66°C (150°F). This means that the collector has to be operating at a temperature difference of at least 66°ΔC (120°ΔF) above outside air temperature. From Figure 5.4, it is seen that an insolation level of 452 cal/cm²min (100 Btu/ft.²hr.) is required to sustain a 120°ΔF. Returning to Figure 5.3, it shows that at about 1500 hours the insolation drops below this level.

Even though storage cannot be charged after 1500 hours, Curve "Q" shows that there is still a substantial amount of heat output from collectors. Most of this heat can go directly to satisfy the demand from the house—which is why storage temperature levels off until collector output drops below demand—but there is more heat than needed by the demand, so some ends up being wasted.

It may be worthwhile to investigate whether or not the heat contained in this overflow could be diverted to a second storage subsystem that is designed to be kept at lower, but still usable temperatures. This second system would only need to be a small fraction of the size of the main storage unit, since it would probably be used only during those periods of weak or failing sunlight when the collector temperature would drop below storage temperature. The heat stored in this smaller storage unit would then be used first, when storage is being discharged, to insure that its temperature is kept low. In this way, it is always held ready to absorb heat during periods of marginal collection. Compared with existing examples of

116

unitary storage system, the utilization of a split, or differential, storage might well extend the capability of the system.

The foregoing is a model of storage and climatic conditions for a fairly good day. For comparison, Figure 5.5 models the effect of bad weather or heavily overcast conditions. This is shown by the curve "CC", or cloud cover.

Cloud cover is a weather bureau measurement and it is reported in tenths of cloud cover, Figure 5.6 graphically presents insolation levels on a bright clear day (sunrise to sunset cloud cover 1/10); a completely overcast day (sunrise to sunset cloud cover 10/10); and a day with a cloud cover of 7/10. These data on cloud cover were collected from the Weather Bureau in Wilmington, Delaware. These data also show that even on a day with 7/10 cloud cover, fully

50% of the insolation for a clear day is received. Although cloud cover is an indicator of poor weather conditions, it should be noted that cloud cover measurements do not necessarily have a direct relationship to the amount of solar energy received.

In Figure 5.5, there is not too much fluctuation in the outside ambient air temperature; to some extent this is an effect of the cloud cover. When there is heavy cloud cover, there is an insulating effect, as if it were a blanket, that keeps the night-time temperatures from dropping too much. It also keeps the day time temperatures from getting much higher than the moderated temperature of the previous night, since the clouds also insulate against the warming effect of the sun. Because the outside air temperature is fairly steady, the demand curve, "D", which is dependent on the ambient temperature, is also steady.

Around 1200 hours, the cloud cover has thinned enough to allow some collection to begin. Storage temperature starts to level out as collector output

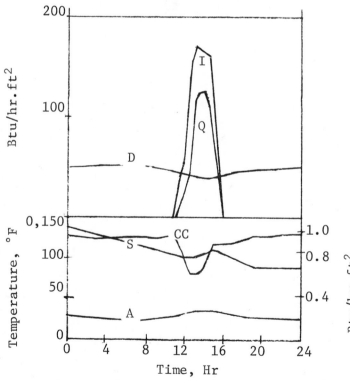

Figure 5.5

DEGREE DAYS and
SOLAR RADIATION

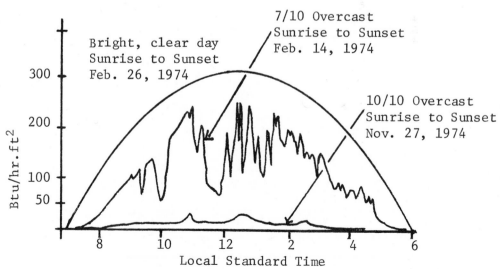

[4]Figure 5.6

takes over the house heating load. Between 1300 and 1500 hours collector output exceeds the immediate demands for space heat. Even though insolation levels are low and the collector temperature is correspondingly low, this output is still able to charge storage, but this is only because the storage temperature has dropped fairly low and is below the temperature of the collectors. At 1500 hours the energy supply curve "Q" drops below the demand curve "D" and storage discharges heat to meet the deficiency. At approximately 1900 hours, storage reaches its minimum operating temperature. The auxiliary heating system takes over and will supply the heating needs until such time as the solar heating system is able to take it over again. The temperature of storage will remain almost fixed at its minimum operating temperature because it will have been well insulated against heat loss.

The above discussion has modeled the dynamics of storage under two very different daily circumstances. But in order to get a better under-standing of the system's overall response, it is important to look carefully at the climate conditions for longer periods of time and under actual load conditions. This is necessary to develop a sense for reasonable expectations of a particular design in a given climate.

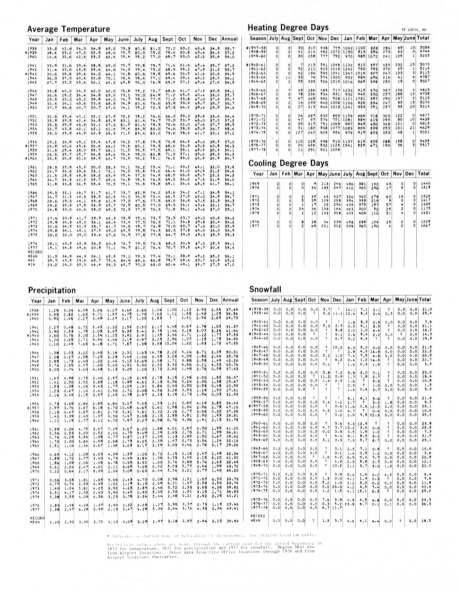

Figure 5.7

Figure 5.8

CLIMATE DATA

For designers to estimate the size of a solar heating or hot water system it is critical that they acquire reasonably accurate and reliable, average weather data for:

DEGREE DAYS and SOLAR RADIATION

Because weather patterns can vary radically from place to place—even over relatively short distances—it is important to try to get the information needed from the closest possible weather stations. The information that is needed can be readily obtained from the National Oceanic and Atmospheric Administration (NOAA) of the U.S. Department of Commerce. Most major cities have National Weather Service Forecasting Stations—generally at airports—that monitor the local weather. The readings from such stations are reported monthly, in "Local Climatological Data" sheets (Figure 5.7), and yearly in an "Annual Summary with Comparative Data" (Figure 5.8).[5]

DEGREE DAYS

The data that is necessary to establish the long-term average monthly degree days is found in the Annual Summary with Comparative Data. Figure 5.9 is that part of the comparative data which presents the monthly heating degree days over many years. To illustrate why long-term averages are used, the average degree days for each month are presented with the highest and lowest monthly degree days for the same period. There **are** considerable differences from year to year.

While the monthly data sheets in any one year are not reliable enough to judge long-term requirements, they can be used to develop a feel for the dynamics of weather demands.

The data presented here are for St. Louis at Lambert Airport, which is really the hinterland for St. Louis City. If a designer were to build a system within the city, it is important that he know that the city is generally several degrees warmer year-round than its outlying areas.[6] This would require some adjustment of the degree day totals so that they more closely represent the actual climate in the St. Louis area. Data kept as yearly total degree days show that Lambert Airport has a yearly total of roughly 4900 DD while the total for the city is only 4500 DD.[7] This means that the monthly averages for Lambert have to be modified so that they are usable as monthly averages for the city of St. Louis. A simple ratio bet-

HEATING DEGREE DAYS

SAINT LOUIS, MISSOURI

Season	July	Aug.	Sept.	Oct.	Nov.	Dec.	Jan.	Feb.	Mar.	Apr.	May	June	Total
1934-35	0	3	73	157	474	1021	989	772	460	411	173	18	4551
1935-36	0	7	46	269	684	1115	1275	1213	528	432	37	6	5612
1936-37	0	0	33	254	701	848	1075	887	752	300	72	6	4928
1937-38	0	0	61	303	690	1035	974	621	348	287	93	0	4412
#1938-39	0	0	42	143	551	931	814	881	557	392	58	1	4370
1939-40	0	0	20	215	640	885	1553	930	696	370	129	0	5438
1940-41	0	0	77	133	678	810	953	930	787	220	42	3	4633
1941-42	0	0	39	174	584	775	1036	890	588	219	109	21	4435
1942-43	0	2	131	237	525	1041	1058	728	788	332	99	2	4943
#1943-44	0	0	70	261	696	1056	892	806	754	367	98	9	5009
1944-45	0	5	27	213	562	1139	1147	834	375	269	198	48	4817
1945-46	0	0	52	250	595	1153	961	693	269	196	115	20	4304
1946-47	0	6	35	145	517	773	871	1022	879	317	125	8	4698
1947-48	0	0	68	83	728	831	1217	916	672	212	75	0	4802
1948-49	0	0	25	282	533	841	1016	809	671	342	36	0	4555
1949-50	0	0	100	199	498	782	841	779	751	440	54	5	4449
1950-51	0	4	37	117	796	1114	996	824	765	427	58	7	5145
1951-52	1	0	56	234	779	965	896	724	693	331	88	0	4767
1952-53	0	0	29	368	572	864	914	693	558	403	82	0	4483
1953-54	0	0	19	169	516	842	992	588	703	162	172	12	4175
1954-55	0	2	14	281	568	869	983	808	622	107	28	26	4308
1955-56	0	0	9	230	692	979	1101	811	615	368	74	12	4891
1956-57	0	0	27	75	600	816	1175	686	664	317	72	0	4432
#1957-58	0	0	30	310	638	745	1042	1100	828	284	69	10	5056
1958-59	0	2	41	214	482	1072	1190	813	594	273	63	0	4744
1959-60	0	0	30	258	777	792	970	985	1072	241	129	1	5255
#1960-61	0	0	7	213	551	1048	1132	810	597	455	232	25	5070
1961-62	0	0	52	211	623	1015	1242	792	783	370	25	1	5114
1962-63	0	0	62	194	595	1041	1347	1016	507	247	103	3	5115
1963-64	0	11	33	74	574	1320	932	884	696	216	41	6	4787
1964-65	0	2	55	333	526	1019	1016	869	938	250	19	0	5027
1965-66	0	7	49	286	489	717	1233	915	570	397	156	4	4823
1966-67	0	0	78	336	534	931	932	940	530	257	188	12	4738
1967-68	3	7	67	268	682	928	1121	1031	587	290	137	2	5123
1968-69	0	2	14	293	640	1008	1106	826	834	247	85	15	5070
1969-70	0	0	27	313	644	1013	1241	893	751	257	55	20	5214
1970-71	0	0	24	287	635	863	1159	866	718	303	122	0	4977
1971-72	0	0	47	97	574	751	1081	884	619	295	80	10	4438
1972-73	2	0	29	317	751	1069	997	849	430	348	121	0	4913
1973-74	0	0	31	182	538	1077							

\# Indicates a break in the data sequence during the year, or season, due to a station move or relocation of instruments.

	Oct	Nov	Dec	Jan	Feb	Mar	Apr
High	368	796	1320	1553	1213	1072	455
Low	74	474	717	814	621	269	107
Avg	224	611	947	1064	854	655	306

Figure 5.9

ween the different degree days is used as a factor to calculate the monthly degree days for St. Louis (Figure 5.10). These figures were generated for St. Louis, but the approach is readily applicable to other climate areas. The important thing is to try to establish reasonably accurate long-term averages for the specific site involved. Correlations of this kind are the most straightforward way to generate the averages when actual readings are not available.

```
St. Louis City:Lambert Airport
        4500:4900              = .92

        Oct   Nov   Dec   Jan   Feb   Mar   Apr
Airport 224   611   947  1064   854   635   306
City    206   562   871   979   786   584   282
```

Figure 5.10

SOLAR RADIATION

The idea of developing ways to correlate data to transform it to usable information for other sites is even more necessary when trying to establish reliable solar radiation data for a particular microclimate. The reason for this is that there are not very many weather stations that record this information, and that there are fewer still that have collected data over enough years to develop reasonably reliable averages. A procedure similar to that for degree days was required to establish data for insolation in St. Louis. It should be noted here that columns 18 and 19 in Figure 5.7, which report hours and percent of possible sunshine, have no direct relationship to the amount of solar energy available. They represent the amount of time there is enough solar energy to activate a solar cell; it does not measure the quantity of energy.

The nearest source of long-term solar radiation data for St. Louis is Columbia, Missouri—about 125 miles away. For all practical purposes, Columbia and St. Louis are at the same latitude; but there are significantly lower levels of pollution in the more rural Columbia. There was very little solar radiation data

available in St. Louis to develop correlations from, so care was taken to try to make it as meaningful as possible.

The St. Louis information that was available was recorded at St. Louis University near downtown St. Louis.[8] The data were in the form of hourly totals of Langleys on a horizontal surface from October 1 to October 10, 1973, inclusive. Each day of data was compared directly with data for the same day in Columbia, Missouri. Figure 5.11 compares the average total radiation received for each hour (solar time) for the two locations. The graph shows the pattern of diurnal variation in solar energy. Each location has four hours of peak radiation, but the St. Louis data is skewed, and occurs an hour earlier. The curves show that, except for the 10:00 a.m. anomaly, Columbia consistently receives more radiation than St. Louis, which shows a similar, but depressed curve.

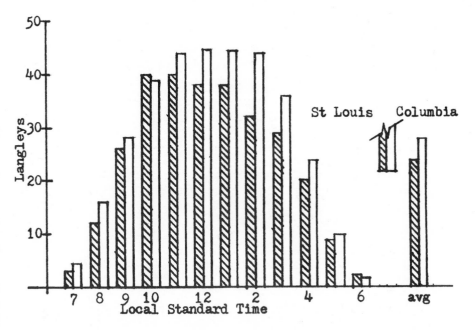

Figure 5.11

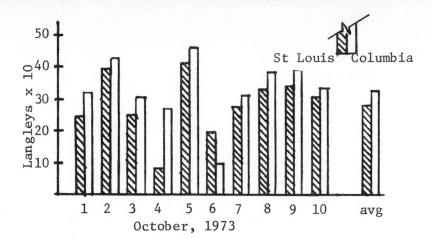

Figure 5.12

Figure 5.12 compares the total radiation received for each day. Again the Columbia data are consistently higher, even as the weather varies through clear and cloudy days. Over this ten day period, St. Louis received an average of 86% of the solar energy that was recorded in Columbia. For this correlation, it was necessary to assume that the instruments were accurately calibrated to the same standard. Because of the way that data were recorded or saved, direct daily data comparison were only possible for this ten day period. Further verification of this .86 correlation was achieved by comparing actual monthly averages of December '72 and January '73 in St. Louis with ten-year, average monthly data (Figure 5.13) for December and

	Solar Radiation			Degree Days	
	Columbia:	St. Louis	%	Average	72-73
Dec '72	167	119	72	947	1069
Jan '73	195	181	93	1064	997

Figure 5.13

January in Columbia. This gave a .72 correlation for December and a .93 correlation for January. It was observed, however, that December '72 had a much worse than average number of degree days. With worse than normal weather, the insolation data could well be expected to be worse than average. By the same token, since January '73 had fewer than the average degree days, its insolation data could be expected to be better than average. These two observations should temper the correlations for December and January and bring them into closer agreement with the October '73 correlation. If this is the case, they tend to validate the .86 correlation of St. Louis radiation data to the long-term, average radiation values for Columbia. Figure 5.14 presents the average daily total of direct and diffuse radiation for Columbia and their corresponding values for St. Louis (based on a factor of .86). This correlation is judged to be reasonably accurate—with a tendency to be on the conservative side.

Although it is possible to develop a correlation on scanty data, it is suggested that the more data like

JAN	FEB	MAR	APR	MAY	JUN	JUL	AUG	SEP	OCT	NOV	DEC	
195	263	342	439	523	585	579	516	396	309	200	167	Columbia (langleys)
168	226	294	378	450	503	498	444	341	266	172	144	St Louis (langleys)
618	834	1084	1392	1658	1855	1836	1636	1256	980	634	530	St Louis (Btu/ft^2)

Average Daily Total of Direct and Diffuse Radiation Incident on a Horizontal Surface
(Langley)(3.687) = Btu/ft^2

[9]**Figure 5.14**

the October '73 period that can be compared, then the more accurate the correlation will be. It is important that the designer attempt such correlations, if only because it will make him much more aware of the microclimate for the design.

COLLECTOR TILT

The data recorded by NOAA on available solar energy (variously: insolation, solar radiation), is the energy that falls on a surface horizontal to the ground plane. The angle of the sun with respect to the ground plane will have a profound effect on the quantity of energy falling on this horizontal plane. As discussed in the section on concentrating collectors, a plane that is kept perpendicular to the sun's rays throughout the day will receive the maximum amount of direct sunlight. Figure 5.15 from Farrington Daniel's **Direct Use of the Sun's Energy** graphically shows the difference in amounts of energy received on surfaces at different tilt angles with respect to the sun. For late October, at 43°N. latitude, a surface normal to the sun receives almost 2.5 times the energy on a surface horizontal to the ground. The time of year and the latitude are specifically pointed out because they are very impor-

tant variables in determining how much extra energy can be received as a function of collector tilt.

For passive solar systems, collector tilt is a moot point. The economies that result from simply using the south wall as the collector more than overcome any losses in efficiency in a technical sense. The overriding equation for efficiency of a system should be:

Energy to Load Per Dollar

which favors the south wall for passive, and even some active solar heating systems.

The horizontal fixed surface and a perpendicular tracking surface are the low and high extremes for solar energy collection (Figure 5.16).

It is generally assumed that the overwhelming majority of residential projects that use active systems

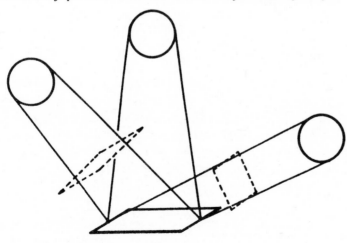

[10]**Figure 5.15**

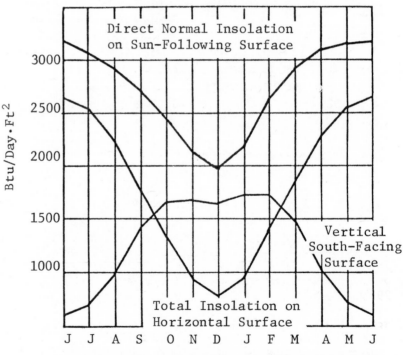

Total Daily Insolation at 40° N Latitude
21st Day of Each Month

[11]**Figure 5.16**

122

will use flat-plate collectors with fixed tilt angles. The **flat-plate collector** is chosen basically because it promises to have lower production costs. A **fixed tilt angle** is selected on the assumption that these units are likely to be roof mounted, and that adjustments for solar declination would create numerous maintenance problems.

What is needed for the FPC's is some reasonable compromise between the fixed horizontal and tracking situations. These flat plates can be tilted up toward the equator to provide some approximation of a surface perpendicular to the **majority** of the daily sun path during the season when heat collection is most critical.

Selecting an optimum tilt requires careful consideration of solar geometry, the proposed uses of a solar heat collection system, and a knowledge of dominant weather patterns within the microclimate of the site. A simple method of selecting the single most effective collector tilt angle for a particular solar application is, therefore, necessary.

Selection of a collector tilt angle is one of the most important decisions to be made concerning an installation. This decision not only gives a solar house its identity and form, but it can also have a significant impact on the cost of the installation. Since the performance of the collector is a function of its tilt, and since the collectors generally represent the most costly component of the installation, significantly lower first costs can be realized if the tilt that is selected provides the best overall performance: the better the performance is, then the less will be the area of collector required to do the job.

It is a well accepted fact that tilting the collector so that it is perpendicular to the incoming solar radiation will yield the best performance at any one time; it is possible to determine graphically the tilt required to produce this perpendicular orientation at a particular time.

Figure 5.17 is an analemma, a basic navigational aide used to correct clock time to actual solar time

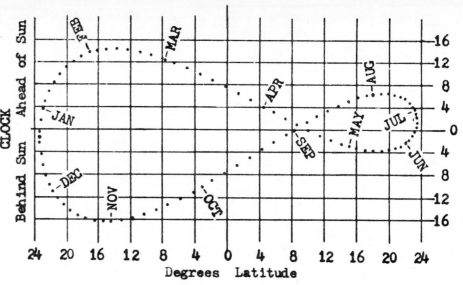

[12]**Figure 5.17**

and to show the latitude where the sun appears to be directly overhead at noon on a given date.

For solar purposes, it allows any day of the year to be easily related to the solar declination, with respect to the equator, for that day. On the analemma, the spring and fall equinoxes correspond to the "O" point on the abcissa. On these two days, the sun is directly over the equator at solar noon.

The latitude at any position on the earth is the angle made by the lines from the center of the earth to that position and to the equator. Since the sun is directly over (0° declination, 90° Altitude) the equator (0° latitude) at solar noon on the equinoxes (Figure 5.18), simple geometry shows that the sun will be perpendicular to any surface that is tilted toward the equator the same number of degrees as the latitude (Figure 5.19). In Figure 5.18 line "AB" is the ray from the center of the earth to the equator, and line "AC" is the ray to the circle of 30° north latitude. The tangent at "C" is representative of the fact that at point "C" the earth is flat. Tilting a line up from "C" toward the equator an amount equal to the degrees of latitude places this line perpendicular to

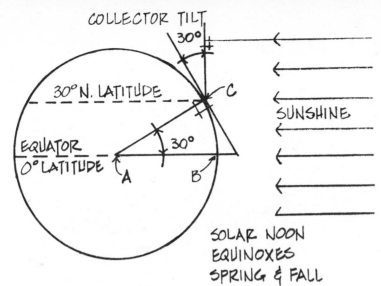

Figure 5.18

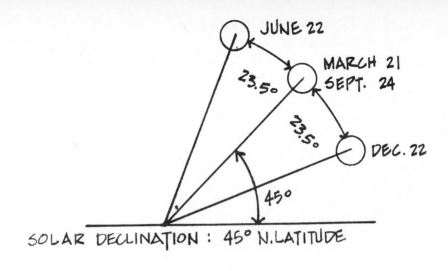

SOLAR DECLINATION : 45° N. LATITUDE

Figure 5.20

incoming rays of sunlight.

Reading sun angle charts for noon at the equinoxes will yield the same information. Figure 5.19 shows the sun's noon angle for 30°, 38°, and 46°. 46°N. latitude, on March 21st, the angle for planes tilted perpendicular to the sun's angle are 30°, 38°, and 46°. They are exactly the same as the latitude. This would be the optimum tilt for collecting solar energy at noon on the equinoxes.

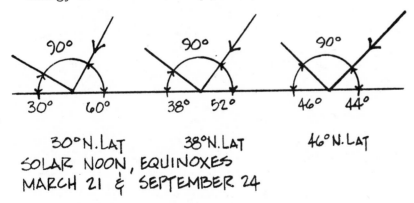

[13]**Figure 5.19**

Since tilting a surface toward the equator an amount equal to its latitude creates a perpendicular configuration at the equinoxes, the seasonal declination of the sun will change with respect to the tilted surface in the same way it changes at the equator. It will move a maximum of 23.5°, plus and minus, from the perpendicular noon position (Figure 5.20). This information allows the construction of Figure 5.21, which substitutes the latitude of the solar installation for "O" on the analemma. This provides a record of how much the solar position at noon deviates from the perpendicular on any given day. Once a determination has been made of the period during which collection is most critical, the collector tilt can be adjusted so that the collectors are perpendicular to the solar noon position during this period. Assuming the collector is already at a tilt angle equal to the latitude, the amount of adjustment (+ or −) can be read directly from the abscissa of Figure 5.21 for any particular period. The tilt adjustments will be positive for winter periods and negative for summer periods.

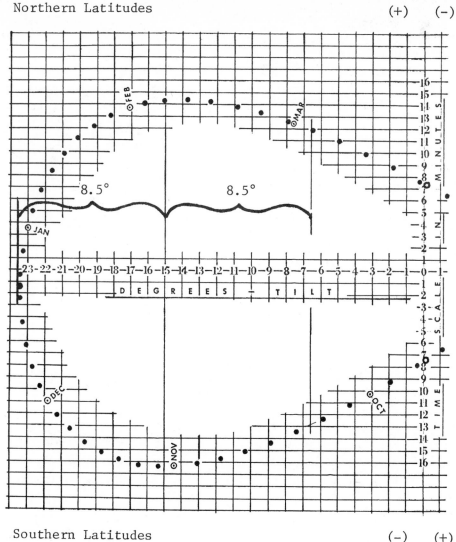

Figure 5.21

In selecting a collector tilt for winter heating, Farrington Daniels suggests that, because more heat is needed in winter than in summer, and because the sun is lower in the sky, more favorable heating performance is achieved if the collector is tilted at an angle 15° greater than the angle of latitude.[14] (For summer cooling, he suggests latitude minus 15 degrees.) Referring to the analemma (Figure 5.21), it

is found that this optimizes solar performance around November 4 and, again, about February 8.

But why latitude plus 15°? If one looks closely, it seems that this optimizes the tilt for the midpoint of a winter's heating season. At latitude plus 15°, the sun's angle at noon on December 23 will be off the perpendicular by 8.5°. Going 8.5° the other direction from latitude plus 15° (Figure 5.21), the same angle of incidence occurs around both October 10th and March 4th. This seems to imply that it is no more important to collect heat in December than it is in mid-October, when winter has really just begun, or in early March, when it is coming to a close. This might be the case if one lived in extremely cold climates, and the Octobers and Marchs were as harsh as December and January. However, this is not the case in most of the United States. It seems that the latitude plus 15° mark was set based on optimizing the solar geometry over a time period called "winter" that had been defined as going from mid-October until March. It does not seem to have been based on nuances in weather or heating needs.

There are several factors that reinforce trying to optimize for more severe weather. There are more hours of sunshine in early and late winter to offset a less than optimum collector orientation (Figure 2.11) and, because the angle of the sun is higher in October and March, more energy reaches the surface of the earth through a shallower cut of atmosphere to compensate for this less advantageous tilt. For the opposite reasons, that the hours of sun are shorter and that the path through the atmosphere is longer, sapping away more energy, the collector tilt should be set to enhance the mid-winter performance when the heat is needed most.

Another reason for not setting the optimums for November 4 and February 8, besides the fact that the demand for heat is comparatively lower than in December and January, is the fact that the solar heating system probably has a limited heat storage capacity. It is very likely—especially if it is optimized for November and February—that storage would often be fully charged. This would mean that, while the collectors were designed to collect the most energy during this time, a lot of heat would have to be vented—wasted—because it just could not be held in an already full storage. This heat "overflow" could not be carried forward to supply heat during more severe weather. So, although there may actually be more insolation collectable over the length of the winter for a tilt of latitude plus 15°, less of this insolation might actually be useable. If the system is designed so that when the collector is at an optimum orientation, all of the available energy can go to load, without losses to overflow, it will be less likely that a system will end up being oversized. The Energy to Load per dollar is maximized.

At some point these considerations of solar collector geometry have to be related to actual system demand. Figure 5.22 is a curve of the average daily degree days per month for St. Louis and is representative of the seasonal heating load. Figure 5.16 is a curve of the maximum energy available. It is for a surface that tracks perpendicular to the sun throughout the day. Since this curve represents the year-round, daily maximums for energy collection, the large dip in the curve is a direct function of the fact that the days become shorter and the fact that the sun must penetrate more miles of atmosphere during the winter months.

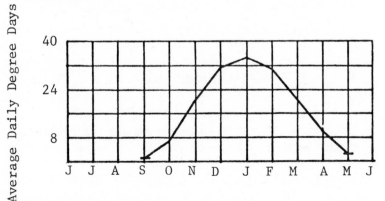

Figure 5.22

Combining these two figures (Figure 5.23) it is apparent that the worst insolation does not coincide with the coldest weather. They are out of phase by a month. From this, it is obvious that, while there is least total insolation in December, it is more critical that the system be optimized for January heat collection.

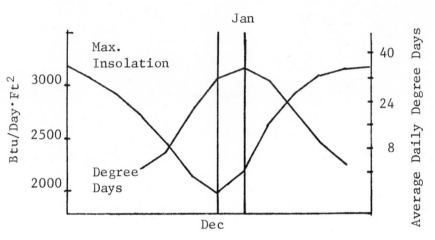

Figure 5.23

A collector tilt can be considered optimum if it is as close to perpendicular as possible to as much of the available energy as possible—during this critical January period. In the literature, optimization is generally carried out with respect to the solar noon angle. However, if a tilt is set to be perpendicular at noon, then there is only one daily maximum; but, if the tilt is set for 10 or 11 a.m. then there will be a second maximum at 1 or 2 p.m. Figure 5.24 shows the daily insolation values for a perpendicular surface by hours for the 21st of each month. From 10 a.m. to 2 p.m. the curves are relatively flat; the differences are relatively small, varying from 9% in December to 3% in March. Since the actual energy levels at these hours are very close, it is more reasonable to consider these additional hours in the optimization process as well as the noon angle.

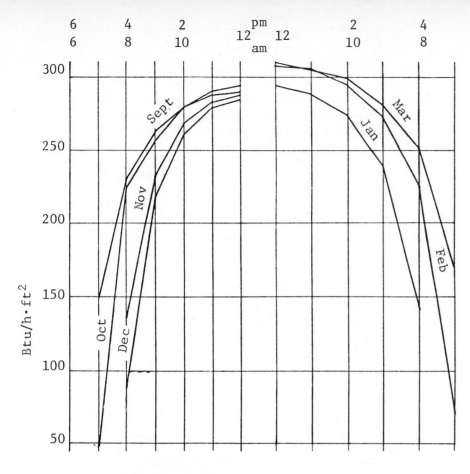

¹⁵**Figure 5.24**

Figures 5.25 and 5.28 are a series of curves for the different hours of the day and their deviation from the perpendicular for different collector tilts. The figures are for 40° N latitude, they are overlayed on the degree day curve for St. Louis from Figure 5.22.

Latitude plus 10° is the tilt that Farrington Daniels suggests should be used for solar hot water heating, which is a constant year-round heating load.[16] For hot water, it is probably best not to orient it specifically for winter. However, Figure 5.25 compares the occurance of optimum periods of collection when the tilt is latitude plus 10° with the dynamics of the winter heating load. It is apparent that latitude plus 10° is

not an optimum for the St. Louis heating season. The optimum peak occuring in early winter is primarily wasted, since it misses the load almost completely. The angles are way off in December and the second peak just barely catches the shoulder of the heating curve. Such a tilt would produce a significant amount of heat overflow early in the winter, but it would not be able to readily handle the heavy loads in late winter.

Figure 5.26 is for latitude plus 15°, the suggested winter optimum according to Farrington Daniels. The peaks for tilt optimum have closed in a little better on the heavier parts of the heating curve than latitude plus 10° did. But the peaks still do not coincide with the most critical part of the winter. It is still slightly off an optimum orientation.

Figure 5.27 is for latitude plus 20°. This particular tilt has pulled the optimum peaks more tightly together, better coinciding with the period of maximum heating load. For most of the heating season the sun angle is off of perpendicular no more than 5° for the hours of 10 a.m. to 2 p.m.

Figure 5.28 is for latitude plus 25°. The peaks are very tightly pulled together. The optimums now occur almost at December 21 (they would be exactly there with a tilt of latitude +23.5°). Although most of

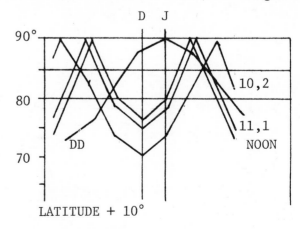

Figure 5.25

the time the sun is again less than 5° off perpendicular, the seasonal coverage of the optimums is narrow and it tails off too quickly under the shoulder of the late winter heating load.

It seems from this, that latitude plus 20° would be optimum orientation for the winter needs of St. Louis, at least as far as this geometrical optimization process is concerned. In general, January will probably be the most severe month in other climates as well; so, latitude plus 20° will probably be optimum for other climates too. Lof and Tybout report, from their computer studies, that the optimum collector tilt is ten to twenty degrees greater than latitude at sites of widely different latitudes. This leaves some room to move around, but it is interesting that for their study they selected latitude plus 15°. One probable reason for this range of tilt angles is the fact that there is data that suggests that until the solar incident angle falls more than 30° away from perpendicular, there is little or no effect on the amount of energy getting through to the absorber (Figure 5.29). Loss of energy becomes very apparent at an angle of 45°. Unfortunately, solar energy is not measured at these varying tilts so that there would be direct, empirical evidence to support one tilt over another. Most of the information available has been calculated through computer simulations using extraterrestrial solar radiation levels. The data generated is only as accurate as the assumptions made that are necessary to modify the data for terrestrial applications. Some of the assumptions generally made are that:

- ground reflectance is 0%; or, that the albedo—reflected solar energy—does not contribute to collectable energy.
- there is an atmospheric clearness factor of 1.0, which substantially reduces the significance of the contribution of diffuse radiation.[18]

Assumptions like this are necessary to generate data in an objective, general way so that variables can be quickly compared and evaluated—as far as the limits of their data suggests. The danger in using such data is in ignoring the differences between test data and the actual environment the collector will be in. This calls for a very specific response and a very subjective interpretation of the extent to which the averaged, general data is applicable.

For example, the above assumptions make the calculated winter data somewhat questionable, especially in northern climates where the snow on the ground has a reflectance of 80% and could make a significant contribution toward solar collection. In

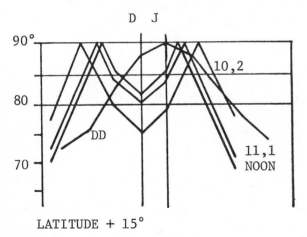

Figure 5.26

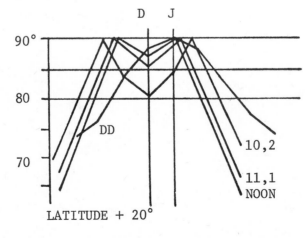

Figure 5.27

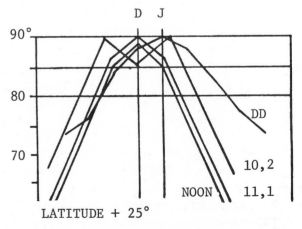

Figure 5.28

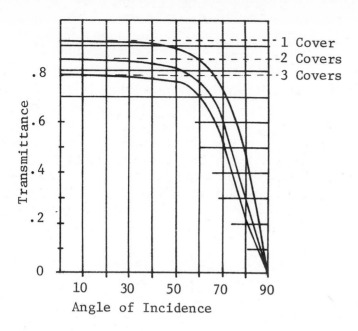

[17]**Figure 5.29**

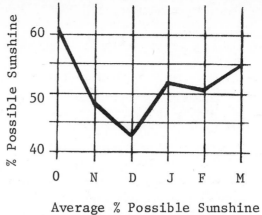

Average % Possible Sunshine
by Months for St. Louis
(1963 - 1972)

[19]**Figure 5.30**

addition, a clearness factor of 1.0 suggests that it is a clear day on which these total daily radiation figures are based. However, during the winter months—at least for St. Louis—the sun only shines approximately 50% of the time (Figure 5.30); and for this 50% the sun was not always shining with a clearness factor of 1.0. As the clarity of the atmosphere decreases, the role that diffuse radiation plays in the collection of solar energy becomes more important.

With the albedo, tilts that are closer to the vertical will be able to take greater advantage of the additional solar energy reflected from the ground. This would suggest that latitude plus 20°, or more even, would perform better than lower tilts. In addressing the question of the effect of diffuse radiation, there does not seem to be enough information available to make any kind of statement about which collector tilts are most favorable for using diffuse radiation. However, intuitively one would expect diffuse radia-

tion to be stronger from the direction of the sun, and that the intensity of diffuse radiation would fall off as the direction of tilt moves away from the sun.

In plotting the more general data from the computer simulations (Figure 5.31), it shows that latitude plus 20° is somewhat better than latitude plus 10°. The information in **ASHRAE, Applications** did not show latitude plus 15°, but it is assumed that it would fall between the two curves, with latitude plus 20° still giving a slightly better performance. This general data structure supports using latitude plus 20° for

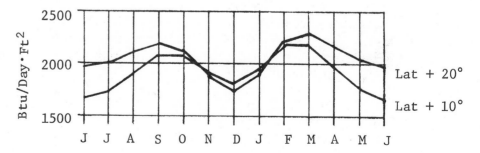

Total Daily Radiation for the
21st of Each Month

[20]**Figure 5.31**

energy collection in January, but it is not overwhelmingly better. However, if this data is properly viewed as a point of departure and such things as albedo are weighed carefully, the sum of these increments of improvement might make it markedly better.

The main point is that one should realize the limits of general data, and that considerations of system orientation have to be firmly referenced in their microclimate. If January is the worst month, then latitude plus 20° seems optimum, provided there are reasonably good levels of insolation during January.

Incidental to the consideration of energy collection, is the fact that steeper tilts will also shed snow and dirt more easily and it has been suggested that steeper tilts be used in areas of violent thunderstorms to cut down on the possibility of hail damage to coverplates.

In general the geometry for collector tilt should be optimized for the period of worst heat loss; tilts optimized for other periods may well collect more energy, but they will also have heavier loss of heat through overflow.

Ultimately, the criteria for judging a system will have to boil down to simple economics: how much usable heat is delivered to load for each dollar invested in the system. On this basis, if one has only a small collector area, orienting for the worst load period might not be best. As long as there will not be overflow, the collectors should be oriented to capture the maximum of energy for the heating season. This might mean optimizing tilt for October and March. Utilizing as much solar energy as possible is ultimately what it is all about.

When space heating is the major concern, larger tilt angles are probably better. The MIT IV solar house, 42°N. latitude, used 60°, which is latitude plus 18°.[21]

If hot water is the sole requirement and requires a year-round optimization, the tilt should be weighted to favor winter somewhat, in order to offset lower insolation levels, but not excessively—latitude plus 10°

would probably be a fairly good tilt.

Solar air conditioning, especially by the absorbtion refrigeration technique, requires high temperatures during the summer. The Colorado State house, designed by Lof to provide 75% heating as well as 75% cooling, has a collector tilt of 45° at a latitude of 40.5° North. This is almost set to the equinoxes, it just slightly favors the winter condition; but the cooling load requires that more attention than usual be given toward a more optimum summer tilt.[22]

MIT's Solar III had a collector tilt of 57° (latitude + 15°) and it provided up to 85% of the heating requirements.[23]

The Solar One House in Delaware (39.5°N. latitude) also has a 45° roof angle. It has the requirement to optimize year-round insolation so that the house can generate as much electricity with its photovoltaic cells as possible.[24]

Michel's house in Montmedy, France at about 50°N. latitude collects over 75% of the heat required.[25] Like most passive designs, it utilizes the south-facing vertical wall as the collector.

To choose one tilt that will be reasonably satisfactory for its intended use is not a simple choice. But such a decision is required if a particular design must be set to one angle only. The real optimum solution would be to design a collector system so that the tilt angle could be adjusted to follow the sun through its seasonal changes (Figure 5.32). However, this may not be economically feasible.

While the decision on a collector tilt is not a light one from a technological point of view, there are other outside factors which must also be considered in a tilt selection. There is a whole question of economics. Weird angles of tilt might prove very costly to build. Aesthetics is also important. FHA requires that neighborhoods of one particular type be maintained in similar styles and forms so that all the properties in the area are maintained at roughly the same market value. A 60° roof in a neighborhood of flat-roofed houses would so deviate from the norm

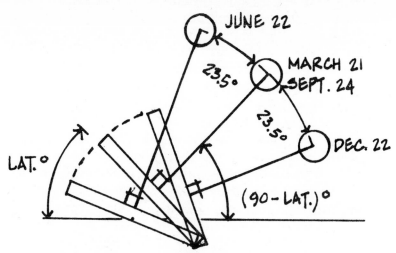

JUNE 22

MARCH 21
SEPT. 24

23.5°

23.5°

DEC. 22

LAT.°

(90-LAT.)°

ADJUSTABLE COLLECTOR FOLLOWS
SEASONAL SOLAR DECLINATION

Figure 5.32

that FHA might well withhold financing in order to preserve the "character" of the neighborhood. Market acceptability is also an important consideration; if solar installations do not look drastically different from the norm, then they might be implemented more quickly.

The discussion of collector tilt has been rather lengthy because selecting a collector tilt is a lot more complex and involved—as well as important—than simply using the latitude plus 15° rule of thumb that occurs so very often in the literature of solar energy. The dynamics of solar availability and winter weather phenomena generally do not lend themselves to simple averaging techniques. An actual application requires individual analysis and attention.

Collector tilt is the one thing that gives a solar house its identity and form; and, from this point of view, it should be a well considered decision.

SYSTEM EFFICIENCIES

From studies done by D. J. Close, it is reasonable to expect that a relatively good air collection system will operate at, or above 50% efficiency with air collection temperatures 56°C (100°F) above ambient air temperatures.[26] In general, water collection systems will operate at lower temperatures, so their efficiencies could be slightly higher than those of air collectors. This efficiency number is determined by measuring the energy content of the coolant (temperature × mass) as it enters the collector, the amount of energy intercepted by the collector and the amount of energy in the coolant as it leaves (Figure 5.33). The net gain in the energy of the coolant as it passes through the collector is compared to the amount of energy intercepted—to yield the efficiency.

From the point where the coolant leaves the collector, the coolant begins to dissipate this energy. There will be thermal loss as it travels to storage, and there will be losses from storage as the heat sits, waiting to be used. In making a judgement as to how much one should expect to lose in these steps it is assumed that 66% (⅔) of the energy that left the col-

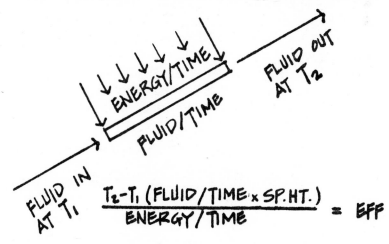

ENERGY/TIME

FLUID OUT
AT T_2

FLUID/TIME

FLUID IN
AT T_1

$$\frac{T_2 - T_1 \; (FLUID/TIME \times SP.HT.)}{ENERGY/TIME} = EFF$$

Figure 5.33

131

lector will ultimately be available for use, through the mechanical system, to heat the house. This 66% has a built in margin of safety in some regards. If storage and all the piping to and from the collector is inside the house, then what might have been heat transport losses is not actually lost, since it will actually heat the house. This 66% factor is also a designer's judgement to accommodate some of the inconsistencies that may arise due to the differences between actual weather conditions and their statistical averages, and to cover the broad assumptions that are necessary to carry through the analysis.

When these individual efficiencies are computed, there is an overall system efficiency—from sunshine to load—of 33%. This means that for every unit of heat delivered to the heating load, three units had to be intercepted by the collector. This system efficiency should be viewed as a conservative, general figure that is reasonably safe to work with, provided one does not ignore the basis for the judgement. To provide a better feel for this figure, the MIT Solar IV House had a 33% overall winter efficiency with collectors that averaged a 45.4% overall, seasonal efficiency and its storage and piping are within the dwelling.[27]

The foregoing discussion establishes the reasonable limits to be expected of an active collector/storage system. They can produce elevated temperatures at rates of 50% efficiency or more, with average winter collection efficiencies of 45%+ —when they are on. Operating time is limited to those periods when the collectors can deliver energy to load (43°C (110°F)) or to storage (49+°C (120+°F)). In severe winter weather, active systems may forfeit the relatively lower levels of insolation (0.9 langleys (200 BTU/H·Ft²) or less) entirely. This is particularly true of the less costly collector assemblies.

Passive systems do not have high-temperature liabilities. Even if the low levels of solar radiation cannot rapidly boost space sensible air temperatures, the radiant aspects of solar energy can directly deposit warmth on the bodies of the occupants. Space Comfort is a function of more than just the air temperature. With the high mean radiant panel temperatures provided by the sun, space air temperatures can fall off drastically and comfort can be sustained. Passive systems have been analyzed for mid-January[28] and late-December[29] performances. The results predict peak winter system efficiencies of 60% and higher. However, definitions of efficiency for Passive Solar systems are currently in flux and will remain that way for some time to come until appropriate ways of evaluating them are developed. A simple transference of Active type analysis methodologies has not proven adequate. Based on mid-winter performances, it may be appropriate to calculate Passive System sizes based on overall, winter seasoned efficiencies of more than 60%. However, since the jury is still out on Passive efficiencies, and since there is little, if no, financial penalty for Passive Solar overdesign (the excess heat can always be vented), a seasonal efficiency of 50% is recommended for sizing calculations. Fifty percent will be used in the following discussions.

SYSTEM CAPABILITIES

Defining the operating limits of the system is basically a task of estimating the point of economic optimum. The single most expensive subsystem is the collectors; so, the main effort should be to make sure that the utilization of the collectors is maximized—getting the most usable heat for the investment. This requires that the area of the collectors be sized so that as little of this energy as possible is wasted through heat overflow. However, at the same time, the system should still make a substantial contribution to the heating load for the most critical part of a season.

Heat input to storage is a function of the collector area and the amount of sunshine. Because the entire system is weather dependent, it is virtually impossible to achieve 100% solar heating. For example, there are long sequences of cloudy days during the winter with eight cloudy days an unusual, but not necessarily infrequent occurrence, in the St. Louis area. If a system were to carry 100% of the heating load, it would have to be able to collect and store enough heat in one day to cover these eight cloudy days. To carry eight days of heating on one day of sunshine would require at least eight times as much collector area as a system designed to carry only one day of heating per day of sunshine. Such a system would still not be able to guarantee 100% solar reliability—there might well be periods of ten cloudy days to contend with. Even a system that carries three or four cloudy days is probably way oversized from an economic standpoint. Some auxiliary heating will always be necessary; and this auxiliary system will have to be the same size as would be necessary if there were no solar heating. This is because the auxiliary system has to be designed to supply enough heat in one hour to meet the maximum expected heat loss for one hour. This one hour maximum is likely to occur when outside conditions are worst—after several very cold, cloudy days—when storage would be exhausted and collectors would be inoperative.

The use of a solar heating system will not provide any savings on the cost of the equipment that would be required to heat the residence conventionally. This means that the cost of the solar installation has to be weighed solely against the fuel savings that result from using solar energy. Based on this, it would be important to size the collector so that the dynamics of the system were such that the collectors would be able to charge the storage unit just enough to carry the heat load for the following night. This way the collector area is used to its maximum all the time. In normal, sunny weather the collectors would be in continuous operation with little or no chance of heat overflow. This would give the lowest cost for each unit of heat collected. If storage for longer periods were required, additional collector area would be needed to charge the system during the preceeding sunny period.

In 1970, Lof and Tybout ran some cost optimization computer programs for various regions in the United States. The program analyzed combinations of different percents of solar performances with auxiliary heating in order to find the lowest annual heating cost per year. The study assumed a solar installation would last 20 years, and was based on fuel prices prior to 1970. The study results are presented in Figure 5.34. These results assumed a heat loss of (25,000 Btu/DD) and it used a collector tilt of latitude plus 15°. The percentage of solar-contributed heat varied from 40-75%.

More recently, with the advent of higher fuel costs and even the uncertainty of whether fuel will be available at any cost, it has been suggested that active solar installations should be designed to carry 60-80% of a total heating needs for a season. It is interesting to note that Lof designed the Colorado State solar house to carry 75% of the heating load with solar power.

Sizing a system to produce 70% of the heat needed during a winter is felt to be very reasonable for all but the most sunless climates. However, this 70% is a recent economic optimum and it may change if the cost of auxiliary fuel increases as much as some predict. This 70% seasonal performance,

	Collector Area (Ft^2)	Output (%solar)	Cost/10^{6}Btu Low	High	100% Electric Heat ($/10^{6}Btu (1967))	($/10^{6}Btu (1962)) Gas--100%--Oil	
Santa Maria	261	75	1.10	1.59	4.36	1.42	1.62
Albuquerque	261	60	1.60	2.32	4.62	0.89	2.07
Phoenix	208	72	2.05	3.09	4.25	0.79	1.60
Omaha	521	47	2.45	2.98	3.24	1.05	1.32
Boston	347	40	2.50	3.02	5.25	1.73	1.76
Charleston	208	55	2.55	3.56	4.22	0.96	1.55
Seattle-Tacoma	521	45	2.60	3.82	2.31	1.83	2.00
Miami	52	70	4.05	4.64	4.90	2.81	1.73

Except for Seattle, where rates are unusually low, least solar costs are less than the cost of all-electric heating.

[30]Figure 5.34

however, does not mean that the system should be sized to carry 70% of the worst month, but that it should carry close to 70% of the heating load for the whole season. For St. Louis, it was found that carrying approximately 50% of the worst month provided a 70% performance over the whole season. While oversizing a system (relative to a 70% system) does manage to increase the amount of solar supplied heat, and reduce auxiliary heat requirements, it adds disproportionately to the cost of each unit of energy produced by the system and generally makes the initial investment in a solar installation a prohibitive, economic burden. The actual size of the various components of the system should be arrived at only after a serious bout with the trade-offs between system cost, expected future fuel costs and the amount of available capital. In view of these uncertainties, perhaps the system should not be designed as a one-shot, completed solar installation. The owner could purchase and install what he can reasonably afford and have the system designed for probable expansion as future economics favor—or even demand it.

SIZING

In order to size a collector and its accompanying storage, it is necessary to transform the foregoing qualitative considerations into quantitative results.

COLLECTORS: As mentioned in the section on insolation, most empirical data have been recorded as radiation on a horizontal surface. Liu and Jordan, using extraterrestrial radiation—eliminating the atmospheric variable—have established multiplying factors that adjust the energy available on a horizontal surface at different latitudes to the amount available on tilted surfaces. The particular tilts used are Latitude and Latitude plus 20°. These factors are for total daily radiation, and the calculations were specifically for the solar angles on the 16th of each

month (Figure 5.35). The Liu and Jordan factors are calculated for the middle of the month; and it is assumed that using these factors will yield a reasonably accurate, average monthly performance for the tilt.

Since these figures are trigonometrically derived based on the solar constant, it is assumed that they are still qualitatively accurate for terrestrial application. Therefore, these factors can be used with actual records of terrestrial, solar insolation on horizontal surfaces. This actual data will have implicity accounted for the atmospheric impedence and the climatic effects on solar energy. Using this technique is about the most direct method available to take advantage of data that have been actually recorded (rather than calculated). However, there are two important qualifications that may effect the available energy levels quantitatively. The first is the fact that the factors were developed on extraterrestrial data where all radiation is direct. It is unclear whether or not the fact that some portion of the total radiation at the earth's surface is diffuse will have a positive or negative affect on amount of energy available at different tilts. The second question is the fact that energy measured on a horizontal surface neglects the albedo, or ground-reflectance. Taking albedo into account would increase the amount of energy available. By neglecting the albedo, it is expected that any error introduced by this technique will still be on the conservative side; that is, actual values would still be higher than calculated values.

Before getting into the actual sizing of the collector it would be helpful to establish the minimum performance required of the system during the critical month of heating. By setting the system to this minimum it should ultimately provide the overall heating performance that is desired for the winter. (The parameters set here will also be used in the discussion of a retrofit solar installation that will follow.)

δ = solar declination
N = total number of hours between sunrise and sunset, hours
H_{oh} = daily radiation on a horizontal surface outside atmosphere of earth, Btu/day-ft^2
R_D = Ratio of daily radiation, H_{ot}, on a surface tilted at β degrees from horizontal toward south to daily radiation, H_{oh}, on a horizontal surface, dimensionless.

[31]Figure 5.35

	δ*	20° North Latitude		$R_D = \frac{H_{ot}}{H_{oh}}$		30° North Latitude		$R_D = \frac{H_{ot}}{H_{oh}}$		40° North Latitude		$R_D = \frac{H_{ot}}{H_{oh}}$		50° North Latitude		$R_D = \frac{H_{ot}}{H_{oh}}$	
		N	H_{oh}	β=20°	β=40°	N	H_{oh}	β=30°	β=50°	N	H_{oh}	β=40°	β=60°	N	H_{oh}	β=50°	β=70°
January	-21°05'	10.93	2415	1.36	1.52	10.29	1906	1.68	1.88	9.50	1364	2.28	2.56	8.38	811	3.56	3.94
February	-21°56'	11.36	2765	1.22	1.28	10.98	2341	1.44	1.52	10.52	1850	1.80	1.90	9.88	1316	2.49	2.62
March	-2°02'	11.90	3138	1.08	1.02	11.64	2857	1.20	1.15	11.77	2489	1.36	1.32	11.68	2046	1.65	1.62
April	9°51'	12.48	3418	1.00	0.83	12.77	3322	1.00	0.84	13.12	3141	1.05	0.90	13.59	2857	1.16	1.00
May	18°55'	12.96	3528	0.92	0.70	13.52	3595	0.87	0.66	14.23	3562	0.86	0.66	15.22	3458	0.90	0.64
June	23°19'	13.20	3547	0.87	0.63	13.92	3691	0.81	0.58	14.83	3761	0.79	0.60	16.12	3739	0.80	0.56
July	21°30'	13.10	3528	0.89	0.66	13.75	3650	0.83	0.62	14.57	3672	0.82	0.64	15.73	3606	0.84	0.62
August	13°59'	12.69	3440	0.95	0.78	13.10	3422	0.93	0.76	13.61	3315	0.96	0.78	14.30	3101	1.02	0.83
September	2°56'	12.14	3233	1.04	0.95	12.23	3038	1.11	1.00	12.33	2736	1.24	1.12	12.47	2360	1.44	1.32
October	-8°36'	11.58	2900	1.17	1.20	11.33	2526	1.36	1.36	11.03	2087	1.62	1.64	10.62	1585	2.10	2.14
November	-18°33'	11.07	2655	1.30	1.44	10.51	2146	1.60	1.76	9.82	1585	2.08	2.24	8.86	1010	3.16	3.32
December	-23°17'	10.80	2308	1.39	1.60	10.08	1781	1.76	1.89	9.18	1224	2.46	2.80	7.83	678	4.04	4.52

* For the 16th day of each month except February. For February, δ is for February 15th.

The system will be required to carry a minimum of 70% of the overall, winter heating load. The collector will be set at latitude + 20°—a 60° tilt. Although St. Louis is actually at 38°45' North Latitude, it is more convenient and probably of no major consequence to use numbers for 40°N. latitude.

The insolation data for a horizontal surface in St. Louis are multiplied by the tilt factors (Figure 5.36) to establish the amount of average daily radiation to be expected on this 60° tilted surface. These insolation values are than normalized to the value for January insolation; that is, each month's daily, average insolation is divided by the average for January. This step just simplifies later calculations, and it shows very clearly the effect collector tilt can have on shifting the point of optimum availability of

	Oct	Nov	Dec	Jan	Feb	Mar	Apr
Average Daily Horizontal Insolation*	265.7	172.0	143.6	167.7	226.2	294.1	377.5
Liu and Jordan Factors (L+20°)	1.64	2.24	2.80	2.56	1.90	1.32	0.90
Average Daily Insolation on 60° Tilt	435.8	385.3	402.1	429.3	429.8	388.2	339.8
Normalized Tilt Insolation	1.014	0.898	0.937	1.000	1.002	0.905	0.792

*St. Louis, Langleys/Day

Figure 5.36

solar energy. This is especially evident when the normalization of horizontal insolation is compared to the normalization after it has been adjusted for a collector tilt of 60° (Figure 5.37).

	Oct	Nov	Dec	Jan	Feb	Mar	Apr
Normalization of Tilted (60°) Insol.	1.014	0.898	0.937	1.000	1.002	0.905	0.792
Normalization of Horizontal Insol.	1.58	1.02	0.86	1.00	1.35	1.75	2.25

Figure 5.37

January was picked for normalization simply because it was the worst month for heat loss, and it would be used to determine the minimum performance of the system under the average, most adverse conditions. Any month could be used for this technique, but it would require that something other than a minimum be found. Using the worst month (generally January) just seems to be a more straightforward approach to a solution.

To determine this minimum an arbitrary first "guess" at the probable percentage of the load carried by the system in January is necessary; for the first iteration, 50% was chosen. This means that with a relative insolation level of 1.0 the system must be able to supply enough heat to offset the loss created by 532 Degree Days (F)—50% of the average monthly total of 1064 DD (F). Now, since all the available solar energy for each of the other months has been normalized to January (January = 1.0), the total number of Heating Degree Days (F) handled by the system is a summation of the monthly insolation normalized factors and the 532 DD (F) load that can be handled in January.

The results are shown in Figure 5.38. As it turns out, the assumption that a system sized to 50% of the January load would produce a 70% seasonal performance was slightly low, providing only 3053 DD (F) of an average total for seven worst months of 4661 DD (F), or 65.4%.

Figure 5.38 also has the second iteration in the process. This time 55% of the January load was the arbitrary figure, and it produced a load carrying capability of 70.8%. From this it will be a simple task, once the Degree Day (C or F) heat loss for the dwelling is established, to size the system. It will be sized to provide 55% of the average January heat load on a daily average basis.

	Oct	Nov	Dec	Jan	Feb	Mar	Apr	
Degree Days*	224	611	947	1064	854	655	306	Total = 4661
Monthly Load January = 50%	539	477	498	532	533	481	421	1st Iteration
Percentage of Monthly Load	100+	78.2	52.5	50.0	62.4	74.5	100+	Overall = 65.4
Monthly Load January = 55%	593	525	548	585	586	529	463	2nd Iteration
Percentage of Monthly Load	100+	86.0	57.8	55.0	68.5	80.7	100+	Overall = 70.8

*DD Based on Lambert Airport, St. Louis, Mo.

Figure 5.38

LOAD: To begin sizing the system, it is important to acquire an accurate picture of the heat loss from the structure that is to be heated. There are well developed methods for calculating this heat loss. Perhaps the single best source is the **ASHRAE Handbook of Fundamentals** and their **Guide and Data Book** which have very comprehensive discussions of heat loss from buildings. However, these methods are structured so that they can be used to size a conventional heating system. A conventional heating system is sized so that it can handle, on an hourly basis, the heat loss when the outside conditions are the severest. For example, McGuiness and Stein suggest that, in St. Louis, the design load be determined by using an outside winter design temperature of −15°C (−5°F). This is an extreme condition that probably only occurs on one or two days or several hours at any one time. This method, calculates the heat loss for one hour, since it is needed only to determine how much the maximum one hour heat output from the conventional system should be. This one hour heat output is the rated capacity—or the size—of the unit.

However, this technique has to be adjusted to use those reasonable, average conditions that a solar installation should be designed to operate within. Figure 5.39 is a schematic set of calculations that illustrate the conventional approach and those adjustments necessary for sizing a solar system.

As in any heat loss analysis, the "U" factors for various surfaces are multiplied by their respective surface areas and these totals are added together with the amount of heat loss expected from air in-

filtration. This sum will yield the heat loss in terms of a factor "A" which will be expressed as kw/°C (Btu/H · °F). "A" is the loss for one hour when there is a temperature difference of 1°C (1°F). This basic factor is common to both sets of calculations.

To size the conventional system, "A" is simply multiplied by the temperature difference between the required inside comfort temperature, 18°C (65°F), and the outside winter design temperature, −15°C d(−5°F).[30] The resulting heat loss will in turn be the required hourly output from the conventional system, or its rated capacity.

For solar estimating purposes, this factor is multiplied by 24 hours to produce the quantity of heat lost in one day for 1°C (1°F). This will be the heat loss for each Degree Day (DD). The system being designed here is supposed to provide for an average of 55% of the monthly degree days for January; therefore, the average number of daily heating degree days that have to be carried by the solar heating system in January is 18.87 DD (F). This is arrived at by dividing the monthly total DD by the number of days and then multiplying by .55.

The actual number of square feet of collector area required is arrived at by multiplying the "Q" (Figure 5.39) by the Monthly Degree Days divided by the number of days in the month, multiplying by the monthly % Solar divided by the seasonal efficiency and dividing by the Average Daily Insolation per square foot as calculated for the tilted surface.

For the above example the equation is:

$$Q_2(18.87)\left(\frac{.55}{.33}\right) \div (429.3) = Q_2(.073) \text{ Ft}^2 \text{ of Collector}$$

This approach allows the designer to quickly assess the impact, in solar equipment cost, of the energy demand of the design. This hopefully, initiates a feedback loop between efforts at conserving energy and efforts at developing new, renewable energy from the environment. Efforts at conserving energy should

$$A = \sum_{i=1}^{n} (U \text{ Factor}_i)(\text{Surface Area}_i)$$

CONVENTIONAL SIZING

$$Q_1 = (A)\binom{\text{Winter Design}}{\text{Temperature}}$$
$$= (\text{Watts/°C})(\Delta°C)$$
$$= (\text{Btu/H·°F})(\Delta°F)$$
$$= \text{Furnace Capacity}$$

SOLAR SIZING

$$Q_2 = (A)(24 \text{ Hours})$$
$$= \text{Watt·H / DD (C)}$$
$$= \text{Btu / DD (F)}$$

COLLECTOR AREA =

$$Q_2 * \left(\frac{\text{Monthly DD}}{\text{Days in Month}}\right)_{\text{Fig. 5.9}} * \left(\frac{\text{Monthly % Solar}}{\text{Seasonal Efficiency}}\right)_{\text{Fig. 5.38}} \div \left(\frac{\text{Avg Daily Insolation}}{\text{for the Month at Selected Tilt Angle}}\right)_{\text{Fig. 5.36}}$$

Active = .33
Passive = .50

Figure 5.39

be enhanced until the cost of conserving that last calorie (BTU) is equal to the cost of developing it from alternative energy sources.

This approach to collector sizing, based on peak load, is more fully developed (for 40°N. latitude) in Appendix I, Normalization Technique, for both Active and Passive Collector systems.

STORAGE: At this point it has been decided that the collectors, with average daily insolation, will be able to supply enough heat to cover 55% of the heat loss for an average number of degree days in January. However, it is assumed that a very good day of sunshine, in a month that averages only half of the possible sunshine hours, could deliver substantially more energy than expected from a day of average insolation. In order to take advantage of all available energy, the storage system should be sized to accommodate this input.

Since the system is sized on the basis of the worst month, it is not unreasonable to assume that at the startup of the day's collection, any energy that had been stored during the previous day will have been depleted while meeting the heating load during the night. With this, it can be assumed that all the heating load for the daytime period will be met by energy coming directly from the collectors and that energy will be put into storage only if there is an excess above the immediate heating demand of the house.

If storage were sized strictly on the basis of averages, then an average day's 9.5 hours of sunshine would supply enough heat for 55% of the average day, or 13.2 hours. With heat going directly to load during the day, storage need only be able to hold 3.7 hours of energy, if the sun produces usable energy for 9.5 hours. This, however, is a very narrow utilization of storage and is almost ludicrous. If such a small storage is actually contemplated, it would probably be more economical to eliminate the storage subsystem entirely and use only the energy that can

be drawn straight off the collectors or stored in the thermal mass of structure. The averages used come from the fact that there are many days of zero input and at least a few days of rather high solar input.

In examining the actual Degree Day records for January (Figure 5.40), one will see that the 55% design criteria will on occasion, be equivalent to 100%, or more, of a specific day's heating requirements. In this case, storage would need to hold, at a minimum, the energy equivalent of 14 (24-9.5) hours times 0.55 of the average day's heating load. However, this is with an average level of insolation. On a warm day, the sun is probably delivering energy well above the average rate. In order to prevent heat overflow (wasted heat) and to keep the system operating at relatively efficient temperature levels, storage would have to be significantly larger than the above calculated 7.7 fully charged hours of storage capacity. Since a relatively good day of sunshine would have at least twice the energy of the average levels of sunshine, a good day could provide enough solar energy to cover 26.4 average hours of January

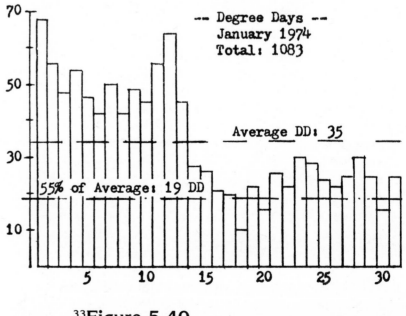

-- Degree Days --
January 1974
Total: 1083

Average DD: 35

55% of Average: 19 DD

³³**Figure 5.40**

heat loss. With 9.5 hours of sunshine, storage now has to be able to accept 17 hours of the average January heating requirements.

Although averages were needed to size collector area, sizing storage requires more specific information on the performance of a "best" day. The size of storage seems to be more reasonable the more specific the data becomes. With good days likely to follow one another, the storage should be able to absorb 34+ hours of the average hours heating load—or about 1.5 average days of heat storage. However, compared to collectors, storage is very inexpensive. One might do well by adding a little extra storage capacity. Although the extra storage might be infrequently used during the worst month, it would probably see extensive usage in less severe months; and it would increase the load carrying capability of the system. This should be done with care, because a storage system that is way oversized for its collector area will not be able to reach the temperatures that it is designed to operate at, since the extra thermal mass would require additional energy to boost it to really useable temperatures. Ideally, the additional storage capacity could be added or deleted as needed, then one would have the ability to take advantage of the really good days and still not make the system sluggish those days of low and moderate levels of insolation. This idea of a differential, or buffer, storage unit should be a topic for future research to discover whether or not, when implemented, this concept provides any significant advantages.

An example has been developed to illustrate the preceding discussion. The basis for the example is a retrofit solar installation. This means that solar collection and storage subsystems are fitted to an existing house. As far as heat loss is concerned, this house will have the characteristics of the general, existing house stock. That is, it will have relatively large expanses of glass, two inches of insulation in frame wall construction, and it will have between 1.5 and 2 air changes per hour due to infiltration. The size of the house has been chosen so that it is the same as the solar house design in Appendix D, which would be new construction. The sizes were correlated so that direct comparisons could be made between an existing housing stock and its potential for solar, and new construction which is designed with the intent of making solar utilization a more economic proposition.

The qualitative and quantitative difference between the two will be the fact that the new solar design takes special heed of heat loss, trying to minimize it. In remodeling the existing dwelling the only major heat loss factor that can be readily and economically reduced is the one for infiltration. This can basically be accomplished by paying careful attention to weatherstripping the perimeters of all openings in the house. Storm windows, which were probably standard equipment before solar utilization was contemplated, serve double duty in reducing infiltration and in lowering the actual heat loss through the glass. A major architectural change in the new solar design is a 33 percent reduction in the amount of glass—down to a level slightly above the FHA minimum requirements. This reduction in glass area cuts down on both the heat loss through glass and losses due to infiltration, since there is less perimeter around fewer openings for air infiltration. A feature that would have helped, but was not incorporated in the design is the use of air-locked entries; placing unheated coat closets or mudrooms between actual entry doorways and the outside environment. This reduces normal infiltration and the heat losses incurred when just opening the door to go in or out.

In the remodeled design, it would probably be much too costly to attempt to reduce heat loss through the walls. New construction, however, has the opportunity to achieve substantial reductions (50% or better) in heat loss through the wall. This reduction would be a result of using 3-3.5 inches of the best available insulation, instead of just the nominal two inches of conventional fiberglass bats—which was the highest quality, industry standard until recently. However, before one worries overly much about insulation requirements, it is important to note how very large heat loss due to infiltration alone is. The best return on one's money would be to attempt to reduce the volume of involuntary air changes (infiltration through cracks) and to install such items as electric ignitions and flue dampers to reduce furnace stack losses (AGA approved, of course).

The particulars for calculating the heat loss are tabulated in Figure 5.41. Those losses are for an existing frame house of approximately 186m² (2000 ft.²). The house has an east-west axis with a south facing roof of 102m² (1100 ft.²) at an angle of 40° from the horizontal.

Figure 5.42 presents the comparative effect on energy required if the air change rate is allowed to vary between 1.5 and 1.0 changes an hour. The collector area to be calculated is the **effective** collector area, the amount of actual working absorber surface. Since this does not include the surface area of the collector housing, approximately 10% is added to account for the casing. This last number establishes the minimum roof area required, if the collectors are to be roof mounted.

RETROFIT	Air Changes	
	1.5/H	1.0/H
Btu/DD Heat Loss	23,620	19,349
Heat Capture Ratio	1:3	1:3
Solar Energy Required/DD	70,860	58,047
55% of Daily Average DD(Jan.)	19	19
Required Btu's Solar Energy	1,346,340	1,102,890

Figure 5.42

	Ft^X	"U" Factor	Btu/H· °F	Btu/H· 70°ΔF	Btu/DD	Btu/ 70DD	Comments
Exterior Wall	2588^2	.082	212	14855	5093	356510	
Ceiling	1045^2	.042	22	1536	526	36820	½ Design ΔT
Slab Edge	145^1	.37	54	3756	1288	90160	
Glass (Storm)	290^2	.56	162	11368	3898	272860	
Volume: A	29664^3	.018	534	37377	12815	897050	1.5 Air Change/Hour
B	19776	.018	356	24917	8544	598010	1.0 Air Change/Hour
Totals: A			984	68892	23620	1,653,400	
B			806	56432	19349	1,354,360	

Figure 5.41

For comparative purposes, the calculations necessary in determining the collector area are executed for collector tilts of 40° and 60°. The calculations use data for the St. Louis microclimate and for both 1.0 and 1.5 air changes an hour. The results of these calculations are presented in Figure 5.43. The corresponding calculations for the new solar design have been brought forward from the appendix. These results are in Figures 5.44 and 5.45.

It was decided that storage would be in water and that it should be able to hold 1.5 days of average demand. But it was also decided that this energy would have to be stored in a 33°ΔC (60°ΔF) temperature difference. This keeps the system operating at reasonably good efficiencies; and, at the same time, if insolation is extremely good, it has the capability of storing the additional energy by boosting the temperature difference even more. If the minimum storage temperature is 26.5°C (80°F), then a 33°ΔC (60°ΔF) rise in temperature tops out storage at 60°C (140°F). In the case of very good insolation, collector temperatures may go as high as 93°C (200°F) for a water system, and higher still if air collection is used.

RETROFIT	Collector Tilt		Comments
	40°	60°	
Total Daily Avg. Insolation (Horiz.)	618	618	Btu/Ft^2 St. Louis
Liu and Jordan Tilt Factor	2.28	2.56	40°N. Lat.
Insolation on Tilted Surface	1409	1582	Btu/Ft^2
Required Effective Collector Area			
1.5 A.C./H	956	850	Ft^2
1.0 A.C./H	783	698	Ft^2
Minimum Roof Area* (at 0.9 Factor)			
1.5 A.C./H	1060	945	Ft^2
1.0 A.C./H	870	775	Ft^2

*Collectors do not have to be roof-mounted.

Figure 5.43

	Ft^x	"U" Factor	Btu/H °F	Btu/H· 70°ΔF	Btu/DD	Btu/ 70DD	Comments
Exterior Wall	2682^2	.035	94	6566	2251	157700	
Ceiling	1045^2	.035	37	1295	878	30723	½ Design ΔT
Slab Edge	145^1		52	3619	1242	87000	
Glass:							
Storm	141^2	.56	79	5516	1891	132384	
Insul.	55^2	.65	36	2513	862	60334	
Volume	14832^3	.018	267	18683	6406	448490	.75 Air Change/Hour
Totals			564	38192	13530	916631	

Figure 5.44

NEW SOLAR	Collector Tilt		
	40°	60°	Comments
Btu/DD Heat Loss	13530		40° N. Lat St. Louis
Heat Capture Ratio	1:3		
Solar Energy Required/DD	40590		Btu's
55% of Daily Average DD (Jan.)	19		
Required Solar Energy/Day (Btus)	771210	771210	Daily Total at 55%
Insolation on Tilted Surface	1409	1582	Btu/ft^2.Day
Required Effective Collector Area	547	487	Ft^2
Minimum roof Area (at 0.9 Factor)	608	541	Ft^2

Figure 5.45

most important step in making solar energy an economic proposition. The less solar equipment required, then the more likely it will be that more people will be able to take advantage of solar energy as an alternate source of energy.

	A.C./H	Mass/ 60°ΔF	Volume Ft^3	Volume m^3	Collector Ft^2
Retrofit	1.0	18,380	290	8.29	870
Retrofit	1.5	22,440	360	10.07	1070
New	.75	12,850	205	5.81	608

Figure 5.46

If need be, this provides an additional 100% capacity between 60-93°C (140-200°F).

Calculations show that storage equal to 1.5 days of average demand (35 DD), at one air change per hours, requires a minimum of 7.5m³ (270 ft.³) of water. This will be more than eight tons of water at a little more than 7570 liters (2000 gallons). Based on the total collector area for a 40° tilt, there is slightly more than 85 kg/m² (17.4 lb/ft²) of water per unit area of collector. This places the storage mass used slightly passed the mid-point of the suggested range for storage mass—as determined earlier in the chapter on Storage.[34]

Figure 5.46 is used to compare the requirements for a retrofit solar house with 1.0 and 1.5 air changes an hour and for a new house that was built with energy conservation and solar utilization in mind. It becomes apparent quite quickly from Figures 5.42 and 5.45 that good, energy-conscious design is the

NOTES

1. The Lof and Tybout study normalized all data to collector area. Values in the preceding figure are "per square foot of collector."

2. The Lof and Tybout study is applicable to both air and water collection systems. For this discussion a "V" grooved absorber with a selective surface has been chosen. Note that the leaving air temperature can be manipulated by accepting lower efficiencies.

3. D. J. Close, "Solar Air Heater," **Solar Energy** (Vol. 7, No. 3, 1963), p. 123. Adapted.

4. T. M. Kuzay, **et. al.**, "Part 3: Insolation and Total System Performance," **Solar One: First Results** (Newark, Delaware: Institute of Energy Conversion, University of Delaware, 1974), pp. 37-39. Compiled from Figures.

5. These weather records are generally available from the local NOAA weather station or from:

   > National Climatic Center
   > Federal Building
   > Ashville, North Carolina 28801

6. W. P. Lowry, "The Climate of the Cities," **Scientific American** (August, 1967.).

7. McGinness and B. Stein, **Mechanical and Electrical Equipment for Buildings** (New York: John Wiley and Sons, Inc., 1971), p. 214.

8. Information on St. Louis insolation was obtained as raw data through Lou Hall at St. Louis University.

9. The Columbia data used were for the years 1961-1971 and are calculated at 2.0 gm cal/min cm² (conversation with David Horner, Columbia, Mo., NOAA Weather Station). Note that both the time and period and Solar Constant are different than the data reported in the Appendix from **The Climate Atlas of the United States**.

10. F. Daniels, **Direct Use of the Sun's Energy** (New York: Ballantine Books, 1974), p. 20.

11. **ASHRAE Handbook and Product Directory: 1974 Applications** (New York: ASHRAE, 1974), p. 59.9 Adapted.

12. The analemma was constructed from a table in **Thermal Environmental Engineering** by J. L. Threlkeld (Englewood Cliffs, New Jersey: Prentice Hall, 1962), p. 286.

13. Declination angles from data by Victor Olgay as recorded in C. G. Ramsey and H. R. Sleeper's, **Architectural Graphic Standards, 6th Edition** (New York: John Wiley and Sons, Inc., 1970), p. 72.

14. F. Daniels, **op. cit.**, p. 41.

15. ASHRAE, **op. cit.**, pp. 59.4-59.5.

16. F. Daniels, **op. cit.**, p. 77.

17. H. C. Hottel and B. B. Woertz, "The Performance of Flat-Plate Solar-Heat Collectors," **Transactions of the ASME,** Feb. 1942, p. 97.

18. ASHRAE, **op. cit.**, p. 59.6.

19. Local Climatological Data, U.S. Department of Commerce, National Oceanic and Atmospheric Administration, St. Louis, Mo. Figure compiled from data.

20. ASHRAE, **op. cit.**, p. 59.9. Adapted.

21. C. D. Engebretson and N. G. Ashar, "Progress in Space Heating with Solar Energy," ASME paper, No. 60 WA-88 (December, 1960).

22. G. O. G. Lof, **et. al.**, "Design and Construction of a Residential Solar Heating and Cooling System," NTIS Doc. No. PB237042 (August, 1974).

23. B. Anderson, **Solar Energy and Shelter Design**, Master of Architecture Thesis, M.I.T., (Jan., 1973), p. 117.

24. K. W. Boer, "Part 1: The Solar One Conversion System," **Solar One: First Results** (Newark, Delaware: Institute of Energy Conversion, University of Delaware, 1974).

25. Department of Housing and Urban Development, Office of International Affairs, "Information Series 24" (August 27, 1973), p. 4.

26. D. J. Close, "Solar Air Heaters," **Solar Energy**, Vol. 7, No. 3 (1963).

27. C. D. Engebretson and N. G. Ashar, **op. cit.**, p. 7.

28. E. Mazria, S. Baker and F. Wessling, "Predicting the Performance of Passive Solar Heated Buildings," **Proceedings of the 2nd National Passive Solar Conference: Volume II**, (March, 1978, Philadelphia), pp. 393-397.

29. M. Schiff, "Direct Gain Passive Solar Design in an Extreme Climate," **Proceedings of the 2nd National Passive Solar Conference: Volume I**, (March, 1978, Philadelphia), pp. 38-42.

30. Lof and Tybout, "Cost of House Heating with Solar Energy," **Solar Energy,** Vol. 14, (1973), pp. 272, 274. Compiled from Figures.

31. B. Y. H. Liu and R. C. Jordan, "Availability of Solar Energy for Flat-Plate Solar Heat Collectors," **ASHRAE, Low Temperature Engineering Applications of Solar Energy** (1967), p. 2.

32. W. J. McGinness and B. Stein, **Mechanical and Electrical Equipment for Buildings** (New York: John Wiley and Sons, Inc., 1971), p. 212, 181. 65°F is the point from which Degree Days are calculated, because buildings seldom require heat until the outside temperature drops below 65°F. St. Louis winter design temperature is −5°F. This means that a conventional furnace is designed to satisfy a ΔT = 70°F.

33. **Local Climatological Data, op. cit.**, Data from January, 1974.

34. G. O. G. Lof and R. A. Tybout, **op. cit.**, p. 276.

   Suggests that 10-15 pounds of water storage per square foot of collector is an economic optimum.

   C. D. Engebretson and N. G. Ashar, **op. cit.**, p. 2.

   The utilization of 1500 gal/608 ft² of collector suggests that approximately 20 lb/ft² collector is optimum for their particular design and climate.

APPENDIX A

MEAN DAILY SOLAR RADIATION, MONTHLY AND ANNUAL

Author's Notes:

The data in this appendix were taken from:

Climate Atlas of the United States
U.S. Department of Commerce
Environmental Science Services Administration
Environmental Data Service
June, 1968

The document is available from either:

Superintendent of Documents
U.S. Government Printing Office
Washington, D.C. 20402

— or —

National Climatic Center
Federal Building
Asheville, North Carolina 28801

A Solar Constant of 1.94 Langleys is used for the International Scale of Pyrheliometry, 1956.

These charts and table are based on all usable solar radiation data, direct and diffuse, measured on a horizontal surface and published in the Monthly Review and Climatological Data National Summary through 1962. All data were measured in, or were reduced to, the International Scale of Pyrheliometry, 1956.

Langley is the unit used to denote one gram calorie per square centimeter (1 langley = 1 gm. cal. cm^{-2}.

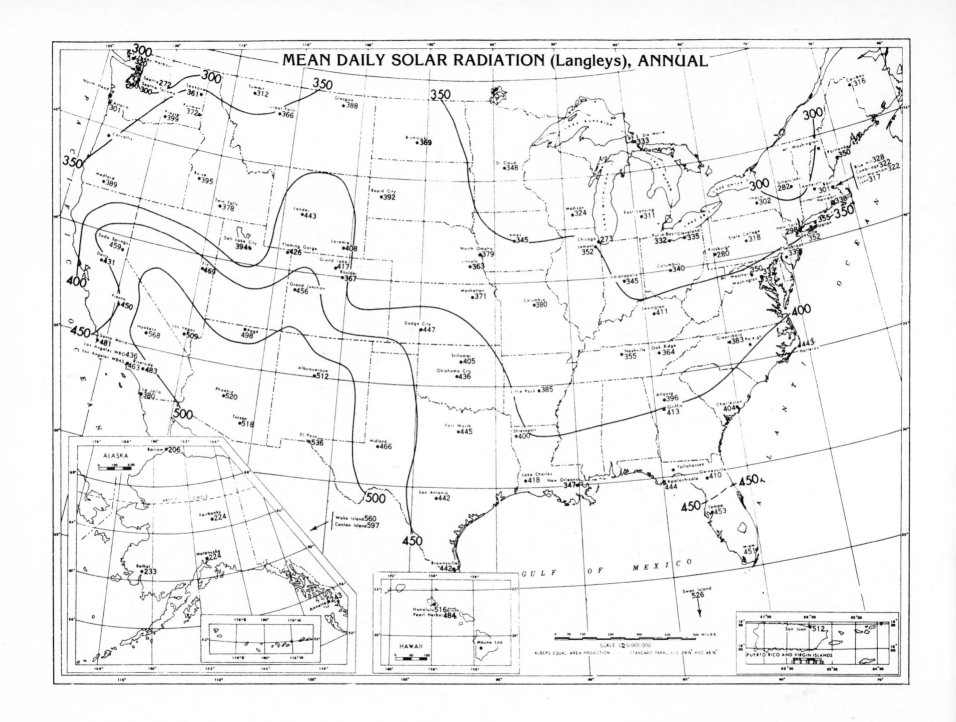

MEAN DAILY SOLAR RADIATION (Langleys), ANNUAL

SCALE 1:20,000,000
ALBERS EQUAL AREA PROJECTION   STANDARD PARALLELS 29½° AND 45½°

ALASKA

HAWAII

PUERTO RICO AND VIRGIN ISLANDS

MEAN DAILY SOLAR RADIATION (Langleys) AND YEARS OF RECORD USED

STATES AND STATIONS	JAN	YRS	FEB	YRS	MAR	YRS	APR	YRS	MAY	YRS	JUNE	YRS	JULY	YRS	AUG	YRS	SEPT	YRS	OCT	YRS	NOV	YRS	DEC	YRS	ANNUAL
ALASKA, Annette	63	6	115	6	236	7	364	7	437	6	438	6	438	6	341	6	258	7	122	7	59	7	41	7	243
Barrow	#		38	8	180	8	380	8	513	8	528	8	429	9	255	10	115	10	41	10	#		#		206
Bethel	38	9	108	10	282	9	444	10	457	10	454	10	376	10	252	10	202	10	115	10	44	9	22	9	233
Fairbanks	16	25	71	27	213	25	376	28	461	28	504	29	434	28	317	29	180	29	82	30	26	26	6	26	224
Matanuska	32	6	92	6	242	4	356	7	436	7	462	6	409	6	314	6	198	6	100	6	38	6	15	7	224
ARIZ., Page	300	2	382	3	526	3	618	2	695	2	707	2	680	3	596	3	516	3	402	3	310	3	243	3	498
Phoenix	301	11	409	11	526	11	638	11	724	11	739	11	658	11	613	11	566	11	449	11	344	11	281	11	520
Tucson	315	5	391	5	540	4	655	5	729	5	699	5	626	6	588	6	570	6	442	6	356	6	305	6	518
ARK., Little Rock	188	9	260	9	353	10	446	9	523	9	559	9	556	8	518	9	439	7	343	8	241	10	187	10	385
CALIFORNIA, Davis	174	18	257	17	390	18	528	18	625	18	694	18	682	18	612	18	493	18	347	19	222	19	148	19	431
Fresno	184	31	289	31	427	31	552	31	647	31	702	32	682	32	621	31	510	31	376	32	250	31	161	32	450
Inyokern (China Lake)	306	11	412	11	562	11	683	11	772	11	819	11	772	11	729	10	635	8	467	9	363	11	300	12	568
LaJolla	244	19	302	18	397	19	457	20	506	19	487	21	497	22	464	22	389	22	320	21	277	20	221	20	380
Los Angeles WBAS	248	10	331	10	470	10	515	10	572	9	596	9	641	9	581	10	503	10	373	10	289	10	241	10	463
Los Angeles WBO	243	9	327	9	436	9	483	9	555	9	584	9	651	9	581	9	500	9	362	10	281	9	234	10	436
Riverside ‡	275	8	367	8	478	9	541	9	623	9	680	9	673	9	618	9	535	9	407	9	319	9	270	9	483
Santa Maria	263	11	346	11	482	11	552	10	635	11	694	11	680	11	613	11	524	11	419	11	313	11	252	11	481
Soda Springs	223	4	316	3	374	4	551	4	615	3	691	4	760	3	681	3	510	3	357	4	248	4	182	3	459
COLO., Boulder	201	5	268	4	401	4	460	4	460	4	525	5	520	5	439	5	412	4	310	4	222	4	182	4	367
Grand Junction	227	9	324	9	434	8	546	8	615	8	708	8	676	8	595	8	514	8	373	10	260	10	212	10	456
Grand Lake (Granby)	212	6	313	7	423	7	512	8	552	8	632	8	600	8	505	7	476	6	361	7	234	6	184	7	417
D.C., Washington (C.O.)	174	3	266	3	344	2	411	2	551	2	494	2	536	2	446	3	375	3	299	3	211	3	166	3	356
American University	158	39	231	39	322	39	398	39	467	39	510	39	496	39	440	38	364	38	278	38	192	39	141	39	333
Silver Hill	177	7	247	6	342	7	438	7	513	7	555	7	511	7	457	7	391	8	293	8	202	7	156	6	357
FLA., Apalachicola	298	10	367	10	441	10	535	10	603	9	578	9	578	9	511	9	456	9	413	10	332	10	262	10	444
Belle Isle	297	10	330	10	412	10	463	10	483	10	464	10	488	11	461	10	400	10	366	11	313	11	291	10	397
Gainesville	267	11	343	10	427	12	517	12	579	12	521	10	488	10	483	8	418	9	347	8	300	10	233	10	410
Miami Airport	349	10	415	9	489	9	540	10	553	10	532	10	532	10	505	10	440	10	384	10	353	10	316	10	451
Tallahassee	274	2	311	2	423	3	499	3	547	3	521	3	508	3	542	2	*		*		292	2	230	2	---
Tampa	327	8	391	8	474	8	539	8	596	8	574	9	534	9	494	9	452	9	400	9	356	9	300	9	453
GA., Atlanta	218	11	290	11	380	11	488	11	533	11	562	11	532	10	508	10	416	10	344	11	268	11	211	11	396
Griffin	234	9	295	9	385	10	522	11	570	11	577	11	556	11	522	11	435	11	368	11	283	11	201	11	413
HAWAII, Honolulu	363	4	422	4	516	4	559	5	617	5	615	5	615	5	612	5	573	5	507	5	426	5	371	5	516
Mauna Loa Obs.	522	2	576	2	680	2	689	3	727	3	*		703	3	642	2	602	2	560	2	504	2	481	3	---
Pearl Harbor	359	5	400	4	487	4	529	5	573	5	566	5	598	5	567	5	539	5	466	5	386	5	343	5	484
IDAHO, Boise	138	10	236	9	342	9	485	9	585	10	636	9	670	10	576	10	460	10	301	11	182	11	124	11	395
Twin Falls	163	20	240	20	355	20	462	21	552	20	592	18	602	20	540	20	432	19	286	20	176	20	131	19	378
ILL., Chicago	96	19	147	19	227	19	331	19	424	19	458	18	473	19	403	18	313	19	207	20	120	20	76	20	273
Lemont	170	6	242	6	340	6	402	6	506	6	553	6	540	6	498	6	398	5	275	5	165	5	138	5	352
IND., Indianapolis	144	10	213	10	316	10	396	10	488	9	543	11	541	10	490	11	405	11	293	11	177	11	132	11	345
IOWA, Ames	174	5	253	5	326	5	403	5	480	5	541	5	436	6	460	6	367	6	274	7	187	7	143	7	345
KANS., Dodge City	255	7	316	7	418	7	528	7	568	7	650	7	642	8	592	9	493	9	380	9	285	10	234	10	447
Manhattan	192	3	264	3	345	4	433	3	527	4	551	4	531	4	526	4	410	4	292	4	227	4	156	4'	371
KY., Lexington	172	9	263	9	357	10	480	10	581	10	628	9	617	10	563	10	494	10	357	9	245	9	174	11	411
LA., Lake Charles	245	11	306	11	397	11	481	11	555	11	591	11	526	11	511	11	449	11	402	11	300	10	250	10	418
New Orleans	214	14	259	14	335	15	412	16	449	14	443	13	417	15	416	15	383	15	357	13	278	13	198	14	347
Shreveport	232	3	292	3	384	3	446	4	558	4	557	4	578	4	528	4	414	4	354	4	254	4	205	4	400
MAINE, Caribou	133	8	231	9	364	8	400	10	476	10	470	10	508	11	448	11	336	11	212	11	111	11	107	9	316
Portland	152	7	235	8	352	7	409	8	514	9	539	9	561	9	488	8	383	7	278	9	157	8	137	9	350
MASS., Amherst	116	2	*		300	2	*		431	2	514	2	*		---		---		---		152	2	124	2	---
Blue Hill	153	27	228	27	319	26	389	26	489	27	510	27	502	26	449	27	354	28	266	28	162	28	135	28	328
Boston	129	16	194	17	290	17	350	17	445	16	483	16	486	16	411	16	334	17	235	16	115	15	135	15	301
Cambridge	153	4	235	3	323	3	400	3	420	3	476	3	482	4	464	4	367	4	253	4	164	4	124	4	322
East Wareham	140	13	218	13	305	12	385	14	452	14	508	14	495	14	436	14	365	13	258	14	163	14	140	13	322
Lynn	118	2	209	2	300	2	394	2	454	2	549	4	528	4	432	3	341	2	241	3	135	3	107	3	317
MICH., East Lansing	121	10	210	11	309	11	359	11	483	10	547	11	540	11	466	11	373	11	255	11	136	11	108	11	311
Sault Ste. Marie	130	10	225	9	356	10	416	10	523	10	557	11	573	11	472	10	322	10	216	9	105	9	96	9	333
MINN., St. Cloud	168	8	260	8	368	8	426	8	496	8	535	8	557	9	486	8	366	8	237	7	146	8	124	8	348
MO., Columbia (C.O.)	173	10	251	10	340	11	434	11	530	11	574	11	574	11	522	10	453	10	322	10	225	10	158	9	380
University of Missouri	166	5	248	6	324	6	429	6	501	6	560	6	583	6	509	6	417	6	324	5	177	5	146	5	365
MONT., Glasgow	154	6	258	8	385	7	466	8	568	8	605	8	645	9	531	10	410	10	267	8	154	8	116	7	388
Great Falls	140	8	232	8	366	9	434	8	528	8	583	8	639	9	532	9	407	10	264	10	154	10	112	10	366
Summit	122	3	162	2	268	3	414	3	462	3	493	3	560	2	510	2	354	2	216	2	102	2	76	2	312
NEBR., Lincoln	188	39	259	39	350	39	416	39	494	40	544	38	568	38	484	38	396	38	296	36	199	40	159	39	363
North Omaha	193	3	299	3	365	3	463	3	516	3	546	4	568	4	519	4	410	4	298	4	204	4	170	4	379

MEAN DAILY SOLAR RADIATION (Langleys) AND YEARS OF RECORD USED

STATES AND STATIONS	JAN	YRS	FEB	YRS	MAR	YRS	APR	YRS	MAY	YRS	JUNE	YRS	JULY	YRS	AUG	YRS	SEPT	YRS	OCT	YRS	NOV	YRS	DEC	YRS	ANNUAL
NEV., Ely	236	7	339	9	468	9	563	9	625	10	712	10	647	11	618	11	518	11	394	10	289	10	218	10	469
Las Vegas	277	11	384	11	519	11	621	11	702	11	748	10	675	11	627	11	551	11	429	11	318	11	258	11	509
N. J., Seabrook	157	8	227	8	318	8	403	8	482	9	527	8	509	8	455	9	385	9	278	7	192	8	140	8	339
N. H., Mt. Washington	117	2	218	2	238	2	*		*		*		---		---		*		*		*		96	2	---
N. Mex., Albuquerque	303	13	386	13	511	13	618	13	686	13	726	13	683	12	626	13	554	14	438	15	334	15	276	14	512
N. Y., Ithaca	116	22	194	21	272	23	334	23	440	24	501	23	513	23	453	23	346	21	231	22	120	23	96	23	302
N. Y. Central Park	130	34	199	34	290	33	369	35	432	35	470	34	459	35	389	35	331	36	242	36	147	36	115	35	298
Sayville	160	11	249	11	335	10	415	10	494	10	565	10	543	10	462	10	385	10	289	10	186	10	142	11	352
Schenectady	130	8	200	9	273	9	338	9	413	9	448	8	441	8	397	8	299	8	218	8	128	8	104	8	282
Upton	155	8	232	8	339	8	428	8	502	8	573	8	543	7	475	7	391	7	293	6	182	7	146	7	355
N. C., Greensboro	200	7	276	9	354	9	469	9	531	10	564	10	544	10	485	10	406	10	322	10	243	10	197	8	383
Hatteras	238	10	317	9	426	8	569	9	635	10	652	10	625	10	562	11	471	11	358	11	282	11	214	11	443
Raleigh	235	3	302	2	*		466	3	494	2	564	2	535	3	476	3	379	3	307	3	235	3	199	3	---
N. D., Bismarck	157	7	250	8	356	6	447	8	550	8	590	9	617	10	516	11	390	11	272	11	161	10	124	10	369
OHIO., Cleveland	125	6	183	6	303	7	286	8	502	8	562	8	562	8	494	8	278	8	289	9	141	9	115	7	335
Columbus	128	7	200	7	297	7	391	7	471	6	562	4	542	5	477	4	422	4	286	4	176	4	129	5	340
Put-in-Bay	126	10	204	9	302	10	386	11	468	11	544	11	561	10	487	10	382	11	275	11	144	11	109	11	332
OKLA., Oklahoma City	251	10	319	10	409	10	494	10	536	10	615	7	610	8	593	8	487	9	377	10	291	9	240	9	436
Stillwater	205	8	289	8	390	9	454	9	504	9	600	10	596	10	545	10	455	11	354	10	269	9	209	8	405
OREG., Astoria	90	7	162	8	270	8	375	8	492	8	469	8	539	8	461	.7	354	7	209	8	111	8	79	8	301
Corvallis	89	2	*		287	3	406	3	517	3	570	3	676	4	558	4	397	4	235	4	144	4	80	4	---
Medford	116	11	215	11	336	11	482	11	592	11	652	11	698	10	605	11	447	11	279	11	149	11	93	11	389
PA., Pittsburgh	94	6	169	5	216	6	317	6	429	6	491	6	497	7	409	6	339	6	207	5	118	6	77	5	280
State College	133	19	201	19	295	20	380	20	456	20	518	20	511	20	444	20	358	20	256	20	149	20	118	20	318
R. I., Newport	155	23	232	22	334	23	405	23	477	23	527	24	513	24	455	24	377	24	271	24	176	24	139	24	338
S. C., Charleston	252	11	314	11	388	11	512	11	551	11	564	11	520	11	501	11	404	11	338	11	286	11	225	11	404
S. D., Rapid City	183	11	277	11	400	11	482	11	532	11	585	11	590	11	541	11	435	11	315	10	204	10	158	10	392
TENN., Nashville	149	18	228	19	322	19	432	19	503	18	551	18	530	17	473	17	403	17	308	19	208	18	150	19	355
Oak Ridge	161	11	239	11	331	11	450	11	518	11	551	11	526	11	478	11	416	11	318	11	213	10	163	11	364
TEXAS, Brownsville	297	10	341	10	402	10	456	11	564	10	610	9	627	8	568	11	475	11	411	11	296	11	263	10	442
El Paso	333	11	430	11	547	10	654	11	714	11	729	11	666	11	640	10	576	11	460	11	372	11	313	11	536
Ft. Worth	250	11	320	11	427	11	488	11	562	11	651	11	613	11	593	11	503	11	403	11	306	11	245	9	445
Midland	283	7	358	8	476	9	550	8	611	8	617	8	608	7	574	8	522	9	396	9	325	8	256	8	466
San Antonio	279	9	347	9	417	9	445	9	541	9	612	9	639	9	585	9	493	10	398	10	295	10	256	8	442
UTAH, Flaming Gorge	238	2	298	2	443	2	522	2	565	2	650	2	599	2	538	3	425	3	352	3	262	3	215	3	426
Salt Lake City	163	8	256	8	354	8	479	8	570	7	621	7	620	6	551	7	446	8	316	8	204	8	146	9	394
VA., Mt. Weather	172	2	274	2	338	2	414	2	508	2	525	3	510	3	430	3	375	3	281	2	202	2	168	2	350
WASH., North Head	*		167	2	257	3	432	2	509	3	487	3	486	3	436	3	321	3	205	3	122	3	77	3	---
Friday Harbor	87	8	157	7	274	8	418	8	514	9	578	10	586	10	507	11	351	8	194	10	102	10	75	8	320
Prosser	117	4	222	4	351	4	521	5	616	4	680	4	707	4	604	4	458	4	274	4	136	4	100	4	399
Pullman	121	4	205	2	304	2	462	2	558	4	653	5	699	5	562	4	410	4	245	5	146	5	96	5	372
University of Washington	67	9	126	9	245	10	364	9	445	10	461	10	496	11	435	10	299	8	170	9	93	9	59	9	272
Seattle-Tacoma	75	9	139	9	265	9	403	9	503	9	511	9	566	9	452	10	324	10	188	10	104	9	64	10	300
Spokane	119	8	204	8	321	8	474	9	563	9	596	9	665	9	536	9	404	10	225	9	131	9	75	7	361
WIS., Madison †	148	46	220	46	313	45	394	47	466	47	514	47	531	47	452	47	348	47	241	47	145	44	115	46	324
WYO., Lander	226	8	324	9	452	9	548	11	587	11	678	11	651	11	586	10	472	8	354	9	239	9	196	9	443
Laramie	216	3	295	3	424	3	508	3	554	3	643	3	606	3	536	3	438	3	324	3	229	3	186	4	408
ISLAND STATIONS																									
Canton Island	588	9	626	7	634	7	604	9	561	9	549	8	550	9	597	9	640	9	651	9	600	8	572	8	597
San Juan, P. R.	404	5	481	4	580	4	622	4	519	5	536	6	536	6	549	6	531	6	460	6	411	6	411	6	512
Swan Island	442	6	496	7	615	6	646	6	625	6	544	8	588	8	591	7	535	8	457	7	394	8	382	8	526
Wake Island	438	7	518	7	577	7	627	7	642	8	656	6	629	7	623	7	587	6	525	7	482	7	421	7	560

NOTES:
* Denotes only one year of data for the month -- no means computed.
--- No data for the month (or incomplete data for the year).
Barrow is in darkness during the winter months.
† Madison data after 1957 not used due to exposure influences.
‡ Riverside data prior to March 1952 not used-instrumental discrepancies.

Langley is the unit used to denote one gram calorie per square centimeter.

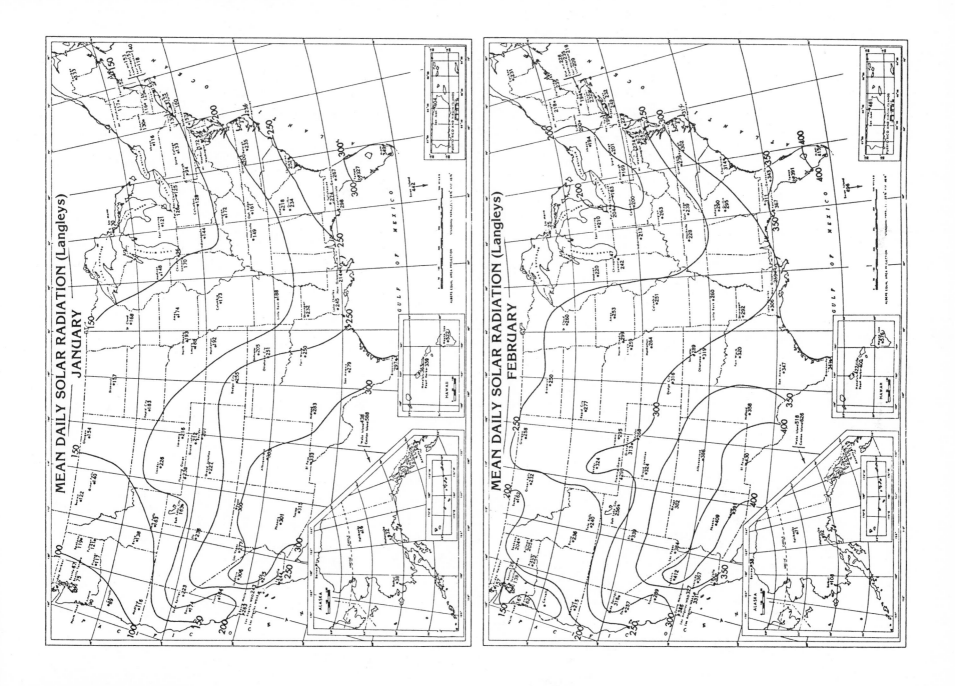

MEAN DAILY SOLAR RADIATION (Langleys)
JANUARY

MEAN DAILY SOLAR RADIATION (Langleys)
FEBRUARY

153

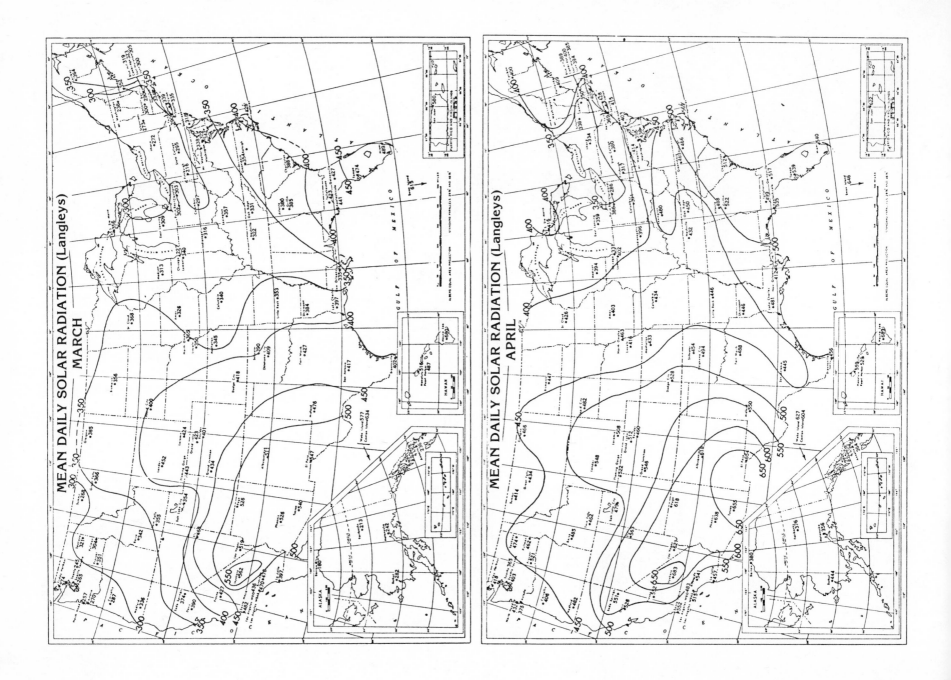

MEAN DAILY SOLAR RADIATION (Langleys)
MARCH

MEAN DAILY SOLAR RADIATION (Langleys)
APRIL

154

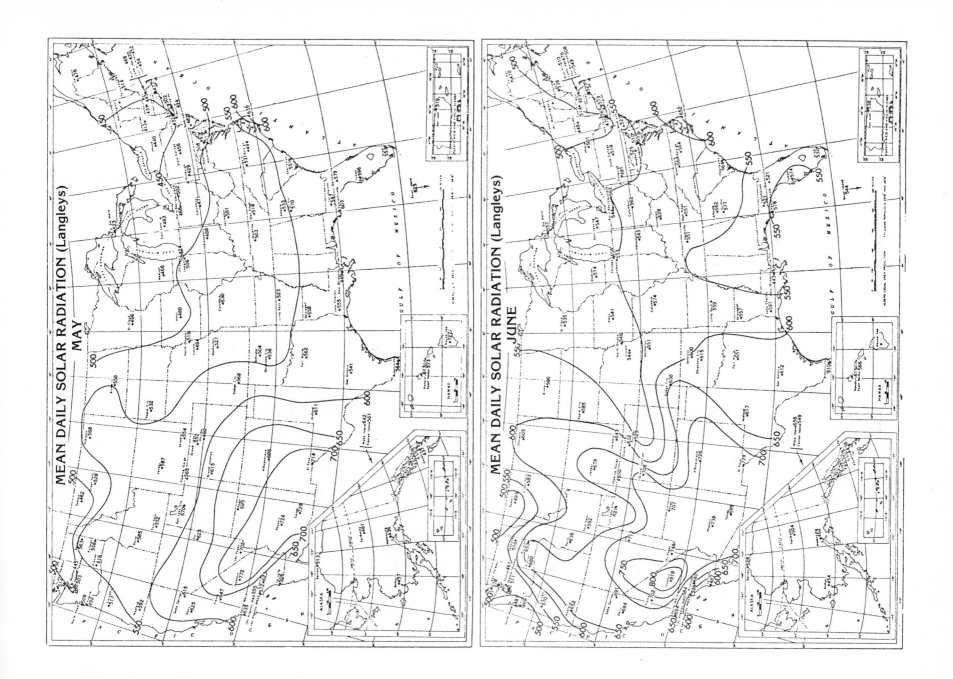

MEAN DAILY SOLAR RADIATION (Langleys)
MAY

MEAN DAILY SOLAR RADIATION (Langleys)
JUNE

155

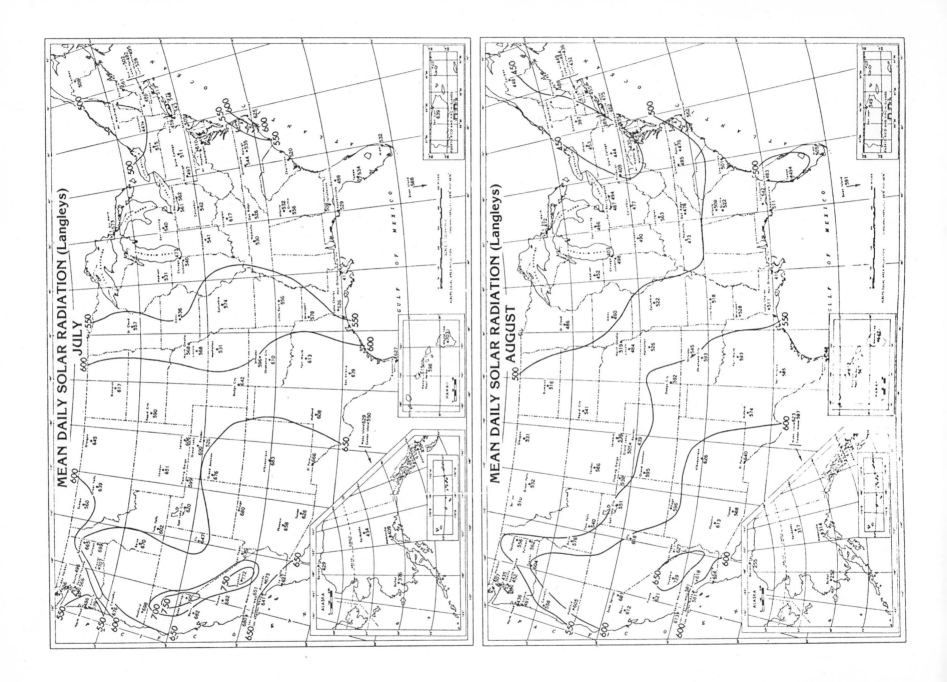

MEAN DAILY SOLAR RADIATION (Langleys)
JULY

MEAN DAILY SOLAR RADIATION (Langleys)
AUGUST

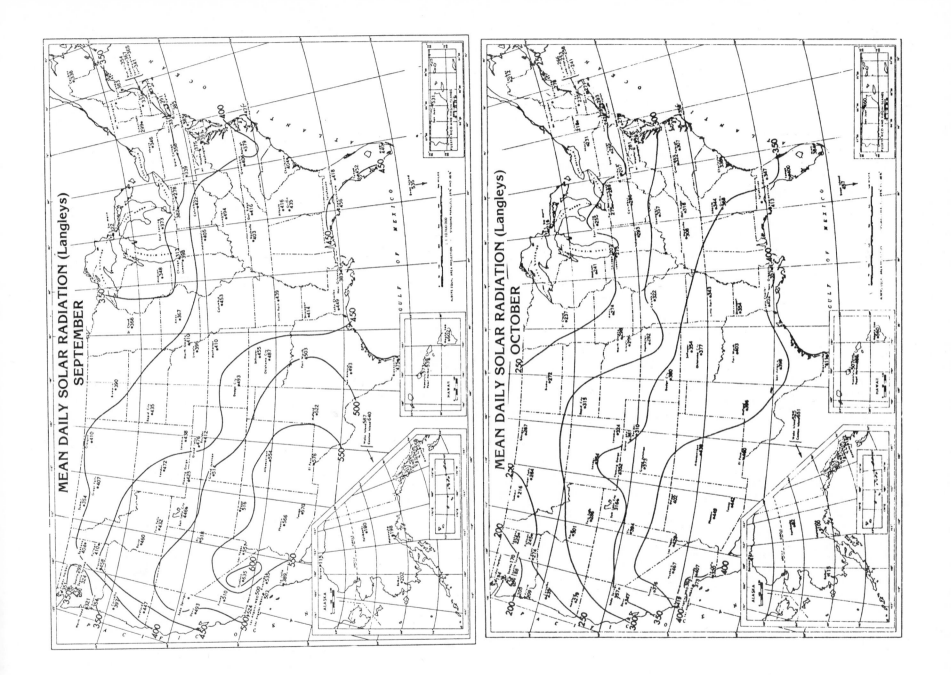

MEAN DAILY SOLAR RADIATION (Langleys)
SEPTEMBER

MEAN DAILY SOLAR RADIATION (Langleys)
OCTOBER

157

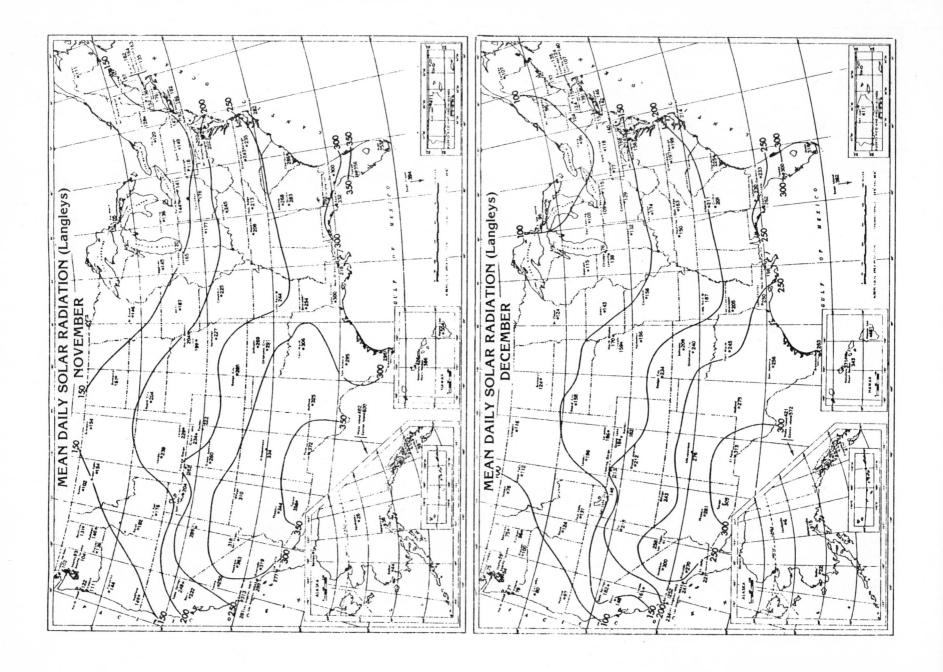

MEAN DAILY SOLAR RADIATION (Langleys)
NOVEMBER

MEAN DAILY SOLAR RADIATION (Langleys)
DECEMBER

APPENDIX B

NORMAL TOTAL HEATING DEGREE DAYS, MONTHLY AND ANNUAL

Author's Notes:

The data in this appendix were taken from:

Climatic Atlas of the United States
U.S. Department of Commerce
Environmental Science Services Administration
Environmental Data Service
June, 1968.

The document is available from either:

Superintendent of Documents
U.S. Government Printing Office
Washington, D.C. 20402

— or —

National Climatic Center
Federal Building
Asheville, North Carolina 28801

One of the most practical of weather statistics is the "heating degree day." First devised some 50 years ago, the degree day system has been in quite general use by the heating industry for more than 30 years.

Heating degree days are the number of degrees the daily average temperature is below 65°. Normally heating is not required in a building when the outdoor average daily temperature is 65°. Heating degree days are determined by subtracting the average daily temperatures below 65° from the base 65°. A day with an average temperature of 50° has 15 heating degree days ($65 - 50 = 15$) while one with an average temperature of 65° or higher has none.

Several characteristics make the degree day figures especially useful. They are cumulative so that the degree day sum for a period of days represents the total heating load for that period. The relationship between degree days and fuel consumption is linear, i.e., doubling the degree days usually doubles the fuel consumption. Comparing normal seasonal degree days in different locations gives a rough estimate of seasonal fuel consumption. For example, it would require roughly 4½ times as much fuel to heat a building in Chicago, Ill., where the mean annual total heating degree days are about 6,200 than to heat a similar building in New Orleans, La., where the annual total heating degree days are around 1,400. Using degree days has the advantage that the consumption ratios are fairly constant, i.e., the fuel consumed per 100 degree days is about the same whether the 100 degree days occur in only 3 or 4 days or are spread over 7 or 8 days.

The rapid adoption of the degree day system paralleled the spread of automatic fuel systems in the 1930's. Since oil and gas are more costly to store than solid fuels, this places a premium on the scheduling of deliveries and the precise evaluation of use rates and peak demands.

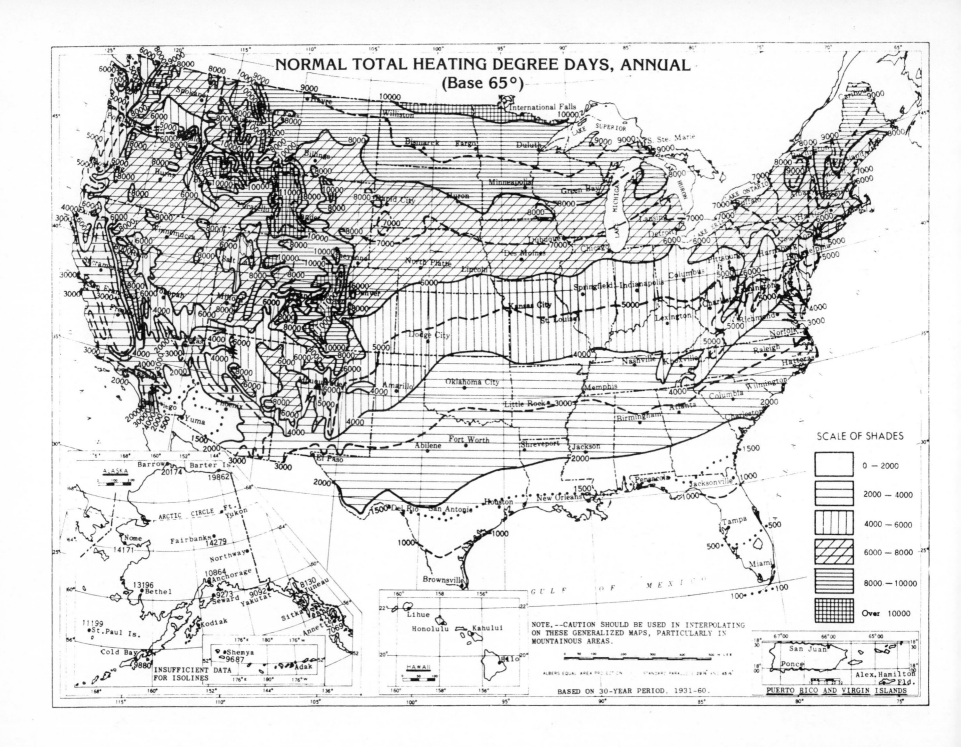

NORMAL TOTAL HEATING DEGREE DAYS, ANNUAL
(Base 65°)

SCALE OF SHADES

	0 — 2000
	2000 — 4000
	4000 — 6000
	6000 — 8000
	8000 — 10000
	Over 10000

NOTE.--CAUTION SHOULD BE USED IN INTERPOLATING ON THESE GENERALIZED MAPS, PARTICULARLY IN MOUNTAINOUS AREAS.

INSUFFICIENT DATA FOR ISOLINES

HAWAII

ALBERS EQUAL AREA PROJECTION   STANDARD PARALLELS 29½° AND 45½°

BASED ON 30-YEAR PERIOD, 1931-60.

PUERTO RICO AND VIRGIN ISLANDS

162

NORMAL TOTAL HEATING DEGREE DAYS
(Base 65°)

STATE AND STATION	JULY	AUG.	SEP.	OCT.	NOV.	DEC.	JAN.	FEB.	MAR.	APR.	MAY	JUNE	ANNUAL
ALA. BIRMINGHAM	0	0	6	93	363	555	592	462	363	108	9	0	2551
HUNTSVILLE	0	0	12	127	426	663	694	557	434	138	19	0	3070
MOBILE	0	0	0	22	213	357	415	300	211	42	0	0	1560
MONTGOMERY	0	0	0	68	330	527	543	417	316	90	0	0	2291
ALASKA ANCHORAGE	245	291	516	930	1284	1572	1631	1316	1293	879	592	315	10864
ANNETTE	242	208	327	567	738	899	949	837	843	648	490	321	7069
BARROW	803	840	1035	1500	1971	2362	2517	2332	2468	1944	1445	957	20174
BARTER IS.	735	775	987	1482	1944	2337	2536	2369	2477	1923	1373	924	19862
BETHEL	319	394	612	1042	1434	1866	1903	1590	1655	1173	806	402	13196
COLD BAY	474	425	525	772	918	1122	1153	1036	1122	951	791	591	9880
CORDOVA	366	391	522	781	1017	1221	1299	1086	1113	864	660	444	9764
FAIRBANKS	171	332	642	1203	1833	2254	2359	1901	1739	1068	555	222	14279
JUNEAU	301	338	483	725	921	1135	1237	1070	1073	810	601	381	9075
KING SALMON	313	322	513	908	1290	1606	1600	1333	1411	966	673	408	11343
KOTZEBUE	381	446	723	1249	1728	2127	2192	1932	2080	1554	1057	636	16105
MCGRATH	208	338	633	1184	1791	2232	2294	1817	1758	1122	648	258	14283
NOME	481	496	693	1094	1455	1820	1879	1666	1770	1314	930	573	14171
SAINT PAUL	605	539	612	862	963	1197	1228	1168	1265	1098	936	726	11199
SHEMYA	577	475	501	784	876	1042	1045	958	1011	885	837	696	9687
YAKUTAT	338	347	474	716	936	1144	1169	1019	1042	840	632	435	9092
ARIZ. FLAGSTAFF	46	68	201	558	867	1073	1169	991	911	651	437	180	7152
PHOENIX	0	0	0	22	234	415	474	328	217	75	0	0	1765
PRESCOTT	0	0	27	245	579	797	865	711	605	360	158	15	4362
TUCSON	0	0	0	25	231	406	471	344	242	75	6	0	1800
WINSLOW	0	0	6	245	711	1008	1054	770	601	291	96	0	4782
YUMA	0	0	0	0	148	319	363	228	130	29	0	0	1217
ARK. FORT SMITH	0	0	12	127	450	704	781	596	456	144	22	0	3292
LITTLE ROCK	0	0	9	127	465	716	756	577	434	126	9	0	3219
TEXARKANA	0	0	0	78	345	561	626	468	350	105	0	0	2533
CALIF. BAKERSFIELD	0	0	0	37	282	502	546	364	267	105	19	0	2122
BISHOP	0	0	42	248	576	797	874	666	539	306	143	36	4227
BLUE CANYON	34	50	120	347	579	766	865	781	791	582	397	195	5507
BURBANK	0	0	6	43	177	301	366	277	239	138	81	18	1646
EUREKA	270	257	258	329	414	499	546	470	505	438	372	285	4643
FRESNO	0	0	0	78	339	558	586	406	319	150	56	0	2492
LONG BEACH	0	0	12	40	156	288	375	297	267	168	90	18	1711
LOS ANGELES	28	22	42	78	180	291	372	302	288	219	158	81	2061
MT. SHASTA	25	34	123	406	696	902	983	784	738	525	347	159	5722
OAKLAND	53	50	45	127	309	481	527	400	353	255	180	90	2870
POINT ARGUELLO	202	186	162	205	291	400	474	392	403	339	298	243	3595
RED BLUFF	0	0	0	53	318	555	605	428	341	168	47	0	2515
SACRAMENTO	0	0	12	81	363	577	614	442	360	216	102	6	2773
SANDBERG	0	0	30	202	480	691	778	661	620	426	264	57	4209
SAN DIEGO	6	0	15	37	123	251	313	249	202	123	84	36	1439
SAN FRANCISCO	81	78	60	143	306	462	508	395	363	279	214	126	3015
SANTA CATALINA	16	0	9	50	165	279	353	308	326	249	192	105	2052
SANTA MARIA	99	93	96	146	270	391	459	370	363	282	233	165	2967
COLO. ALAMOSA	65	99	279	639	1065	1420	1476	1162	1020	696	440	168	8529
COLORADO SPRINGS	9	25	132	456	825	1032	1128	938	893	582	319	84	6423
DENVER	6	9	117	428	819	1035	1132	938	887	558	288	66	6283
GRAND JUNCTION	0	0	30	313	786	1113	1209	907	729	387	146	21	5641
PUEBLO	0	0	54	326	750	986	1085	871	772	429	174	15	5462
CONN. BRIDGEPORT	0	0	66	307	615	986	1079	966	853	510	208	27	5617
HARDFORT	0	6	99	372	711	1119	1209	1061	899	495	177	24	6172
NEW HAVEN	0	12	87	347	648	1011	1097	991	871	543	245	45	5897
DEL. WILMINGTON	0	0	51	270	588	927	980	874	735	387	112	6	4930
FLA. APALACHICOLA	0	0	0	16	153	319	347	260	180	33	0	0	1308
DAYTONA BEACH	0	0	0	0	75	211	248	190	140	15	0	0	879
FORT MYERS	0	0	0	0	24	109	146	101	62	0	0	0	442
JACKSONVILLE	0	0	0	12	144	310	332	246	174	21	0	0	1239
KEY WEST	0	0	0	0	0	28	40	31	9	0	0	0	108
LAKELAND	0	0	0	0	57	164	195	146	99	0	0	0	661
MIAMI BEACH	0	0	0	0	0	40	56	36	9	0	0	0	141
ORLANDO	0	0	0	0	72	198	220	165	105	6	0	0	766
PENSACOLA	0	0	0	19	195	353	400	277	183	36	0	0	1463
TALLAHASSEE	0	0	0	28	198	360	375	286	202	36	0	0	1485
TAMPA	0	0	0	0	60	171	202	148	102	0	0	0	683
WEST PALM BEACH	0	0	0	0	6	65	87	64	31	0	0	0	253
GA. ATHENS	0	0	12	115	405	632	642	529	431	141	22	0	2929
ATLANTA	0	0	18	127	414	626	639	529	437	168	25	0	2983
AUGUSTA	0	0	0	78	333	552	549	445	350	90	0	0	2397
COLUMBUS	0	0	0	87	333	543	552	434	338	96	0	0	2383

NORMAL TOTAL HEATING DEGREE DAYS
(Base 65°)

STATE AND STATION	JULY	AUG.	SEP.	OCT.	NOV.	DEC.	JAN.	FEB.	MAR.	APR.	MAY	JUNE	ANNUAL
MACON	0	0	0	71	297	502	505	403	295	63	0	0	2136
ROME	0	0	24	161	474	701	710	577	468	177	34	0	3326
SAVANNAH	0	0	0	47	246	437	437	353	254	45	0	0	1819
THOMASVILLE	0	0	0	25	198	366	394	305	208	33	0	0	1529
IDAHO BOISE	0	0	132	415	792	1017	1113	854	722	438	245	81	5809
IDAHO FALLS 46W	16	34	270	623	1056	1370	1538	1249	1085	651	391	192	8475
IDAHO FALLS 42NW	16	40	282	648	1107	1432	1600	1291	1107	657	388	192	8760
LEWISTON	0	0	123	403	756	933	1063	815	694	426	239	90	5542
POCATELLO	0	0	172	493	900	1166	1324	1058	905	555	319	141	7033
ILL. CAIRO	0	0	36	164	513	791	856	680	539	195	47	0	3821
CHICAGO	0	0	81	326	753	1113	1209	1044	890	480	211	48	6155
MOLINE	0	9	99	335	774	1181	1314	1100	918	450	189	39	6408
PEORIA	0	6	87	326	759	1113	1218	1025	849	426	183	33	6025
ROCKFORD	6	9	114	400	837	1221	1333	1137	961	516	236	60	6830
SPRINGFIELD	0	0	72	291	696	1023	1135	935	769	354	136	18	5429
IND. EVANSVILLE	0	0	66	220	606	896	955	767	620	237	68	0	4435
FORT WAYNE	0	9	105	378	783	1135	1178	1028	890	471	189	39	6205
INDIANAPOLIS	0	0	90	316	723	1051	1113	949	809	432	177	39	5699
SOUTH BEND	0	6	111	372	777	1125	1221	1070	933	525	239	60	6439
IOWA Burlington	0	0	93	322	768	1135	1259	1042	859	426	177	33	6114
DES MOINES	0	9	99	363	837	1231	1398	1165	967	489	211	39	6808
DUBUQUE	12	31	156	450	906	1287	1420	1204	1026	546	260	78	7376
SIOUX CITY	0	9	108	369	867	1240	1435	1198	989	483	214	39	6951
WATERLOO	12	19	138	428	909	1296	1460	1221	1023	531	229	54	7320
KANS. CONCORDIA	0	0	57	276	705	1023	1163	935	781	372	149	18	5479
DODGE CITY	0	0	33	251	666	939	1051	840	719	354	124	9	4986
GOODLAND	0	6	81	381	810	1073	1166	955	884	507	236	42	6141
TOPEKA	0	0	57	270	672	980	1122	893	722	330	124	12	5182
WICHITA	0	0	33	229	618	905	1023	804	645	270	87	6	4620
KY. COVINGTON	0	0	75	291	669	983	1035	893	756	390	149	24	5265
LEXINGTON	0	0	54	239	609	902	946	818	685	325	105	0	4683
LOUISVILLE	0	0	54	248	609	890	930	818	682	315	105	9	4660
LA. ALEXANDRIA	0	0	0	56	273	431	471	361	260	69	0	0	1921
BATON ROUGE	0	0	0	31	216	369	409	294	208	33	0	0	1560
BURRWOOD	0	0	0	0	96	214	298	218	171	27	0	0	1024
LAKE CHARLES	0	0	0	19	210	341	381	274	195	39	0	0	1459
NEW ORLEANS	0	0	0	19	192	322	363	258	192	39	0	0	1385
SHREVEPORT	0	0	0	47	297	477	552	426	304	81	0	0	2184
MAINE CARIBOU	78	115	336	682	1044	1535	1690	1470	1308	858	468	183	9767
PORTLAND	12	53	195	508	807	1215	1339	1182	1042	675	372	111	7511
MD. BALTIMORE	0	0	48	264	585	905	936	820	679	327	90	0	4654
FREDERICK	0	0	66	307	624	955	995	876	741	384	127	12	5087
MASS. BLUE HILL OBSY	0	22	108	381	690	1085	1178	1053	936	579	267	69	6368
BOSTON	0	9	60	316	603	983	1088	972	846	513	208	36	5634
NANTUCKET	12	22	93	332	573	896	992	941	896	621	384	129	5891
PITTSFIELD	25	59	219	524	831	1231	1339	1196	1063	660	326	105	7578
WORCESTER	6	34	147	450	774	1172	1271	1123	998	612	304	78	6969
MICH. ALPENA	68	105	273	580	912	1268	1404	1299	1218	777	446	156	8506
DETROIT (CITY)	0	0	87	360	738	1088	1181	1058	936	522	220	42	6232
ESCANABA	59	87	243	539	924	1293	1445	1296	1203	777	456	159	8481
FLINT	16	40	159	465	843	1212	1330	1198	1066	639	319	90	7377
GRAND RAPIDS	9	28	135	434	804	1147	1259	1134	1011	579	279	75	6894
LANSING	6	22	138	431	813	1163	1262	1142	1011	579	273	69	6909
MARQUETTE	59	81	240	527	936	1268	1411	1268	1187	771	468	177	8393
MUSKEGON	12	28	120	400	762	1088	1209	1100	995	594	310	78	6696
SAULT STE. MARIE	96	105	279	580	951	1367	1525	1380	1277	810	477	201	9048
MINN. DULUTH	71	109	330	632	1131	1581	1745	1518	1355	840	490	198	10000
INTERNATIONAL FALLS	71	112	363	701	1236	1724	1919	1621	1414	828	443	174	10606
MINNEAPOLIS	22	31	189	505	1014	1454	1631	1380	1166	621	288	81	8382
ROCHESTER	25	34	186	474	1005	1438	1593	1366	1150	630	301	93	8295
SAINT CLOUD	28	47	225	549	1065	1500	1702	1445	1221	666	326	105	8879
MISS. JACKSON	0	0	0	65	315	502	546	414	310	87	0	0	2239
MERIDIAN	0	0	0	81	339	518	543	417	310	81	0	0	2289
VICKSBURG	0	0	0	53	279	462	512	384	282	69	0	0	2041
MO. COLUMBIA	0	0	54	251	651	967	1076	874	716	324	121	12	5046
KANSAS	0	0	39	220	612	905	1032	818	682	294	109	0	4711
ST. JOSEPH	0	6	60	285	708	1039	1172	949	769	348	133	15	5484
ST. LOUIS	0	0	60	251	627	936	1026	848	704	312	121	15	4900
SPRINGFIELD	0	0	45	223	600	877	973	781	660	291	105	6	4561
MONT. BILLINGS	6	15	186	487	897	1135	1296	1100	970	570	285	102	7049
GLASGOW	31	47	270	608	1104	1466	1711	1439	1187	648	335	150	8996
GREAT FALLS	28	53	258	543	921	1169	1349	1154	1063	642	384	186	7750

NORMAL TOTAL HEATING DEGREE DAYS
(Base 65°)

STATE AND STATION	JULY	AUG.	SEP.	OCT.	NOV.	DEC.	JAN.	FEB.	MAR.	APR.	MAY	JUNE	ANNUAL
HAVRE	28	53	306	595	1065	1367	1584	1364	1181	657	338	162	8700
HELENA	31	59	294	601	1002	1265	1438	1170	1042	651	381	195	8129
KALISPELL	50	99	321	654	1020	1240	1401	1134	1029	639	397	207	8191
MILES CITY	6	6	174	502	972	1296	1504	1252	1057	579	276	99	7723
MISSOULA	34	74	303	651	1035	1287	1420	1120	970	621	391	219	8125
NEBR. GRAND ISLAND	0	6	108	381	834	1172	1314	1089	908	462	211	45	6530
LINCOLN	0	6	75	301	726	1066	1237	1016	834	402	171	30	5864
NORFOLK	9	0	111	397	873	1234	1414	1179	983	498	233	48	6979
NORTH PLATTE	0	6	123	440	885	1166	1271	1039	930	519	248	57	6684
OMAHA	0	12	105	357	828	1175	1355	1126	939	465	208	42	6612
SCOTTSBLUFF	0	0	138	459	876	1128	1231	1008	921	552	285	75	6673
VALENTINE	9	12	165	493	942	1237	1395	1176	1045	579	288	84	7425
NEV. ELKO	9	34	225	561	924	1197	1314	1036	911	621	409	192	7433
ELY	28	43	234	592	939	1184	1308	1075	977	672	456	225	7733
LAS VEGAS	0	0	0	78	387	617	688	487	335	111	6	0	2709
RENO	43	87	204	490	801	1026	1073	823	729	510	357	189	6332
WINNEMUCCA	0	34	210	536	876	1091	1172	916	837	573	363	153	6761
N. H. CONCORD	6	50	177	505	822	1240	1358	1184	1032	636	298	75	7383
MT. WASH. OBSY.	493	536	720	1057	1341	1742	1820	1663	1652	1260	930	603	13817
N. J. ATLANTIC CITY	0	0	39	251	549	880	936	848	741	420	133	15	4812
NEWARK	0	0	30	248	573	921	983	876	729	381	118	0	4859
TRENTON	0	0	57	264	576	924	989	885	753	399	121	12	4980
N. MEX. ALBUQUERQUE	0	0	12	229	642	868	930	703	595	288	81	0	4348
CLAYTON	0	6	66	310	699	899	986	812	747	429	183	21	5158
RATON	9	28	126	431	825	1048	1116	904	834	543	301	63	6228
ROSWELL	0	0	18	202	573	806	840	641	481	201	31	0	3793
SILVER CITY	0	0	6	183	525	729	791	605	518	261	87	0	3705
N. Y. ALBANY	0	19	138	440	777	1194	1311	1156	992	564	239	45	6875
BINGHAMTON (AP)	22	65	201	471	810	1184	1277	1154	1045	645	313	99	7286
BINGHAMTON (PO)	0	28	141	406	732	1107	1190	1081	949	543	229	45	6451
BUFFALO	19	37	141	440	777	1156	1256	1145	1039	645	329	78	7062
CENTRAL PARK	0	0	30	233	540	902	986	885	760	408	118	9	4871
J. F. KENNEDY INTL	0	0	36	248	564	933	1029	935	815	480	167	12	5219
LAGUARDIA	0	0	27	223	528	887	973	879	750	414	124	6	4811
ROCHESTER	9	31	126	415	747	1125	1234	1123	1014	597	279	48	6748
SCHENECTADY	0	22	123	422	756	1159	1283	1131	970	543	211	30	6650
SYRACUSE	6	28	132	415	744	1153	1271	1140	1004	570	248	45	6756
N.C. ASHEVILLE	0	0	48	245	555	775	784	683	592	273	87	0	4042
CAPE HATTERAS	0	0	0	78	273	521	580	518	440	177	25	0	2612
CHARLOTTE	0	0	6	124	438	691	691	582	481	156	22	0	3191
GREENSBORO	0	0	33	192	513	778	784	672	552	234	47	0	3805
RALEIGH	0	0	21	164	450	716	725	616	487	180	34	0	3393
WILMINGTON	0	0	0	74	291	521	546	462	357	96	0	0	2347
WINSTON SALEM	0	0	21	171	483	747	753	652	524	207	37	0	3595
N. DAK. BISMARCK	34	28	222	577	1083	1463	1708	1442	1203	645	329	117	8851
DEVILS LAKE	40	53	273	642	1191	1634	1872	1579	1345	753	381	138	9901
FARGO	28	37	219	574	1107	1569	1789	1520	1262	690	332	99	9226
WILLISTON	31	43	261	601	1122	1513	1758	1473	1262	681	357	141	9243
OHIO AKRON	0	9	96	381	726	1070	1138	1016	871	489	202	39	6037
CINCINNATI	0	0	54	248	612	921	970	837	701	336	118	9	4806
CLEVELAND	9	25	105	384	738	1088	1159	1047	918	552	260	66	6351
COLUMBUS	0	6	84	347	714	1039	1088	949	809	426	171	27	5660
DAYTON	0	6	78	310	696	1045	1097	955	809	429	167	30	5622
MANSFIELD	9	22	114	397	768	1110	1169	1042	924	543	245	60	6403
SANDUSKY	0	6	66	313	684	1032	1107	991	868	495	198	36	5796
TOLEDO	0	16	117	406	792	1138	1200	1056	924	543	242	60	6494
YOUNGSTOWN	6	19	120	412	771	1104	1169	1047	921	540	248	60	6417
OKLA. OKLAHOMA CITY	0	0	15	164	498	766	868	664	527	189	34	0	3725
TULSA	0	0	18	158	522	787	893	683	539	213	47	0	3860
OREG. ASTORIA	146	130	210	375	561	679	753	622	636	480	363	231	5186
BURNS	12	37	210	515	867	1113	1246	988	856	570	366	177	6957
EUGENE	34	34	129	366	585	719	803	627	589	426	279	135	4726
MEACHAM	84	124	288	580	918	1091	1209	1005	983	726	527	339	7874
MEDFORD	0	0	78	372	678	871	918	697	642	432	242	78	5008
PENDLETON	0	0	111	350	711	884	1017	773	617	396	205	63	5127
PORTLAND	25	28	114	335	597	735	825	644	586	396	245	105	4635
ROSEBURG	22	16	105	329	567	713	766	608	570	405	267	123	4491
SALEM	37	31	111	338	594	729	822	647	611	417	273	144	4754
SEXTON SUMMIT	81	81	171	443	666	874	958	809	818	609	465	279	6254
PA. ALLENTOWN	0	0	90	353	693	1045	1116	1002	849	471	167	24	5810
ERIE	0	25	102	391	714	1063	1169	1081	973	585	288	60	6451
HARRISBURG	0	0	63	298	648	992	1045	907	766	396	124	12	5251

NORMAL TOTAL HEATING DEGREE DAYS
(Base 65°)

STATE AND STATION	JULY	AUG.	SEP.	OCT.	NOV.	DEC.	JAN.	FEB.	MAR.	APR.	MAY	JUNE	ANNUAL
PHILADELPHIA	0	0	60	291	621	964	1014	890	744	390	115	12	5101
PITTSBURGH	0	9	105	375	726	1063	1119	1002	874	480	195	39	5987
READING	0	0	54	257	597	939	1001	885	735	372	105	0	4945
SCRANTON	0	19	132	434	762	1104	1156	1028	893	498	195	33	6254
WILLIAMSPORT	0	9	111	375	717	1073	1122	1002	856	468	177	24	5934
R. I. BLOCK IS.	0	16	78	307	594	902	1020	955	877	612	344	99	5804
PROVIDENCE	0	16	96	372	660	1023	1110	988	868	534	236	51	5954
S. C. CHARLESTON	0	0	0	59	282	471	487	389	291	54	0	0	2033
COLUMBIA	0	0	0	84	345	577	570	470	357	81	0	0	2484
FLORENCE	0	0	0	78	315	552	552	459	347	84	0	0	2387
GREENVILLE	0	0	0	112	387	636	648	535	434	120	12	0	2884
SPARTANBURG	0	0	15	130	417	667	663	560	453	144	25	0	3074
S. DAK. HURON	9	12	165	508	1014	1432	1628	1355	1125	600	288	87	8223
RAPID CITY	22	12	165	481	897	1172	1333	1145	1051	615	326	126	7345
SIOUX FALLS	19	25	168	462	972	1361	1544	1285	1082	573	270	78	7839
TENN. BRISTOL	0	0	51	236	573	828	828	700	598	261	68	0	4143
CHATTANOOGA	0	0	18	143	468	698	722	577	453	150	25	0	3254
KNOXVILLE	0	0	30	171	489	725	732	613	493	198	43	0	3494
MEMPHIS	0	0	18	130	447	698	729	585	456	147	22	0	3232
NASHVILLE	0	0	30	158	495	732	778	644	512	189	40	0	3578
OAK RIDGE (CO)	0	0	39	192	531	772	778	669	552	228	56	0	3817
TEX. ABILENE	0	0	0	99	366	586	642	470	347	114	0	0	2624
AMARILLO	0	0	18	205	570	797	877	664	546	252	56	0	3985
AUSTIN	0	0	0	31	225	388	468	325	223	51	0	0	1711
BROWNSVILLE	0	0	0	0	66	149	205	106	74	0	0	0	600
CORPUS CHRISTI	0	0	0	0	120	220	291	174	109	0	0	0	914
DALLAS	0	0	0	62	321	524	601	440	319	90	6	0	2363
EL PASO	0	0	0	84	414	648	685	445	319	105	0	0	2700
FORT WORTH	0	0	0	65	324	536	614	448	319	99	0	0	2405
GALVESTON	0	0	0	0	138	270	350	258	189	30	0	0	1235
HOUSTON	0	0	0	6	183	307	384	288	192	36	0	0	1396
LAREDO	0	0	0	0	105	217	267	134	74	0	0	0	797
LUBBOCK	0	0	18	174	513	744	800	613	484	201	31	0	3578
MIDLAND	0	0	0	87	381	592	651	468	322	90	0	0	2591
PORT ARTHUR	0	0	0	22	207	329	384	274	192	39	0	0	1447
SAN ANGELO	0	0	0	68	318	536	567	412	288	66	0	0	2255
SAN ANTONIO	0	0	0	31	207	363	428	286	195	39	0	0	1549
VICTORIA	0	0	0	6	150	270	344	230	152	21	0	0	1173
WACO	0	0	0	43	270	456	536	389	270	66	0	0	2030
WICHITA FALLS	0	0	0	99	381	632	698	518	378	120	6	0	2832
UTAH MILFORD	0	0	99	443	867	1141	1252	988	822	519	279	87	6497
SALT LAKE CITY	0	0	81	419	849	1082	1172	910	763	459	233	84	6052
WENDOVER	0	0	48	372	822	1091	1178	902	729	408	177	51	5778
VT. BURLINGTON	28	65	207	539	891	1349	1513	1333	1187	714	353	90	8269
VA. CAPE HENRY	0	0	0	112	360	645	694	633	536	246	53	0	3279
LYNCHBURG	0	0	51	223	540	822	849	731	605	267	78	0	4166
NORFOLK	0	0	0	136	408	698	738	655	533	216	37	0	3421
RICHMOND	0	0	36	214	495	784	815	703	546	219	53	0	3865
ROANOKE	0	0	51	229	549	825	834	722	614	261	65	0	4150
WASH. NAT'L. AP.	0	0	33	217	519	834	871	762	626	288	74	0	4224
WASH. OLYMPIA	68	71	198	422	636	753	834	675	645	450	307	177	5236
SEATTLE	50	47	129	329	543	657	738	599	577	396	242	117	4424
SEATTLE BOEING	34	40	147	384	624	763	831	655	608	411	242	99	4838
SEATTLE TACOMA	56	62	162	391	633	750	828	678	657	474	295	159	5145
SPOKANE	9	25	168	493	879	1082	1231	980	834	531	288	135	6655
STAMPEDE PASS	273	291	393	701	1008	1178	1287	1075	1085	855	654	483	9283
TATOOSH IS.	295	279	306	406	534	639	713	613	645	525	431	333	5719
WALLA WALLA	0	0	87	310	681	843	986	745	589	342	177	45	4805
YAKIMA	0	12	144	450	828	1039	1163	868	713	435	220	69	5941
W. VA. CHARLESTON	0	0	63	254	591	865	880	770	648	300	96	9	4476
ELKINS	9	25	135	400	729	992	1008	896	791	444	198	48	5675
HUNTINGTON	0	0	63	257	585	856	880	764	636	294	99	12	4446
PARKERSBURG	0	0	60	264	606	905	942	826	691	339	115	6	4754
WIS. GREEN BAY	28	50	174	484	924	1333	1494	1313	1141	654	335	99	8029
LA CROSSE	12	19	153	437	924	1339	1504	1277	1070	540	245	69	7589
MADISON	25	40	174	474	930	1330	1473	1274	1113	618	310	102	7863
MILWAUKEE	43	47	174	471	876	1252	1376	1193	1054	642	372	135	7635
WYO. CASPER	6	16	192	524	942	1169	1290	1084	1020	657	381	129	7410
CHEYENNE	19	31	210	543	924	1101	1228	1056	1011	672	381	102	7278
LANDER	6	19	204	555	1020	1299	1417	1145	1017	654	381	153	7870
SHERIAN	25	31	219	539	948	1200	1355	1154	1054	642	366	150	7683

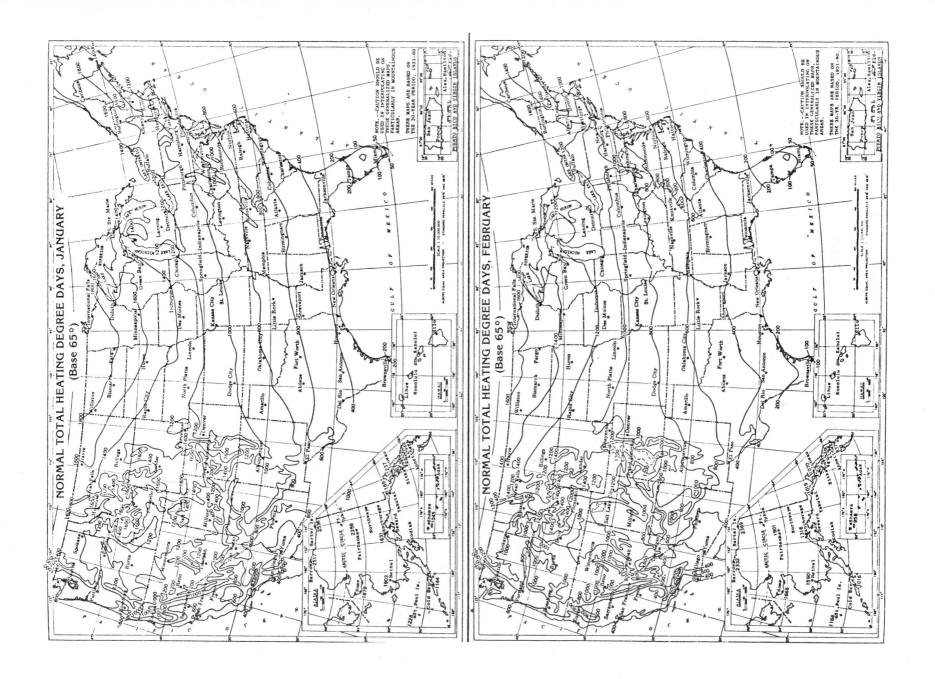

NORMAL TOTAL HEATING DEGREE DAYS, JANUARY
(Base 65°)

NORMAL TOTAL HEATING DEGREE DAYS, FEBRUARY
(Base 65°)

165

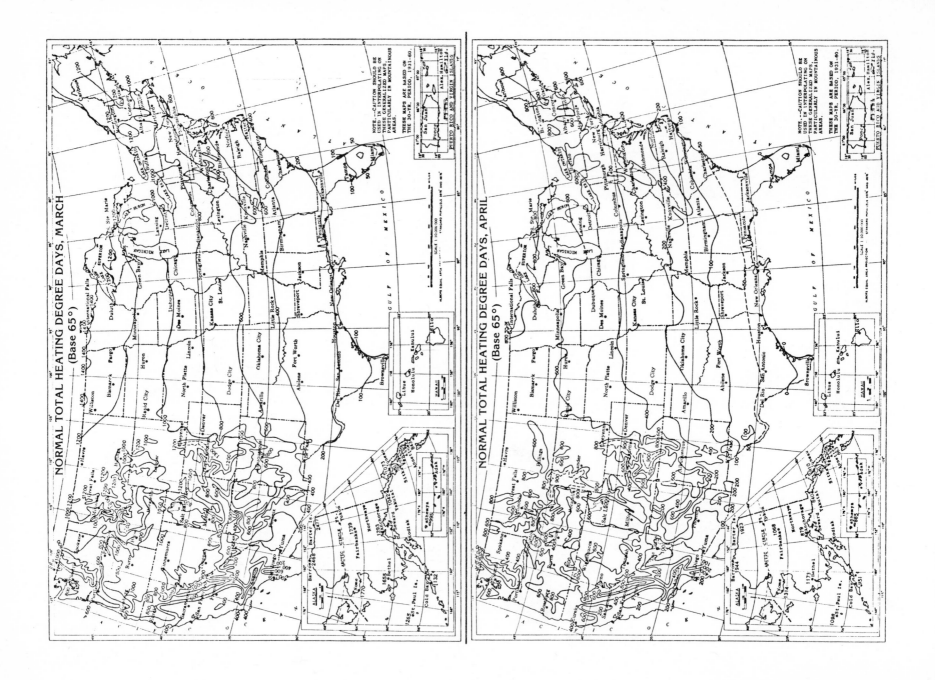

NORMAL TOTAL HEATING DEGREE DAYS, MARCH (Base 65°)

NORMAL TOTAL HEATING DEGREE DAYS, APRIL (Base 65°)

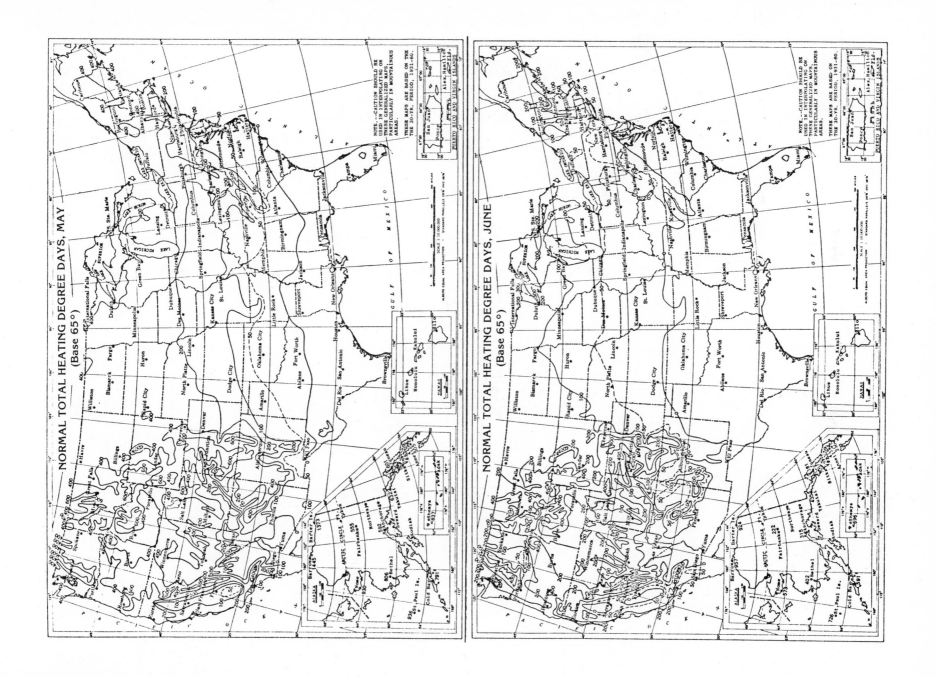

NORMAL TOTAL HEATING DEGREE DAYS, MAY (Base 65°)

NORMAL TOTAL HEATING DEGREE DAYS, JUNE (Base 65°)

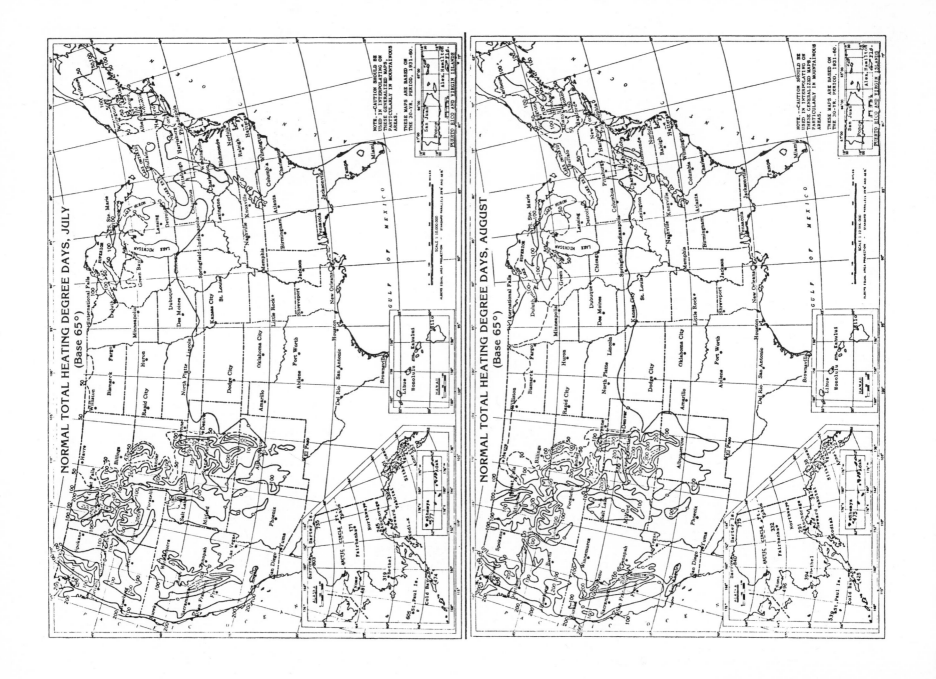

NORMAL TOTAL HEATING DEGREE DAYS, JULY
(Base 65°)

NORMAL TOTAL HEATING DEGREE DAYS, AUGUST
(Base 65°)

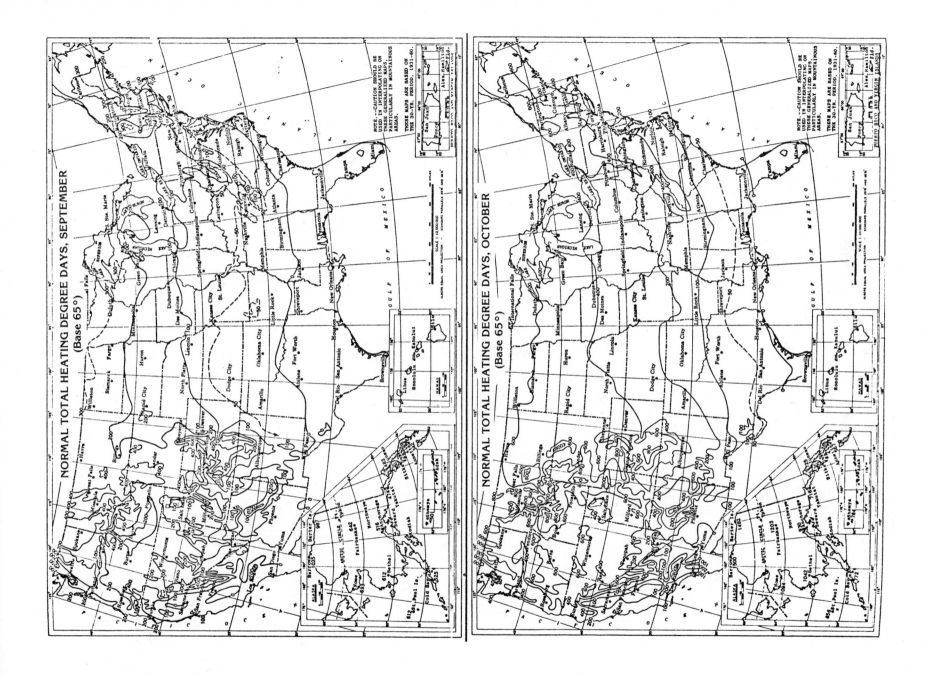

NORMAL TOTAL HEATING DEGREE DAYS, SEPTEMBER
(Base 65°)

NORMAL TOTAL HEATING DEGREE DAYS, OCTOBER
(Base 65°)

NOTE.—CAUTION SHOULD BE USED IN INTERPOLATING ON THESE GENERALIZED MAPS, PARTICULARLY IN MOUNTAINOUS AREAS.

THESE MAPS ARE BASED ON THE 30-YR. PERIOD, 1931-60.

169

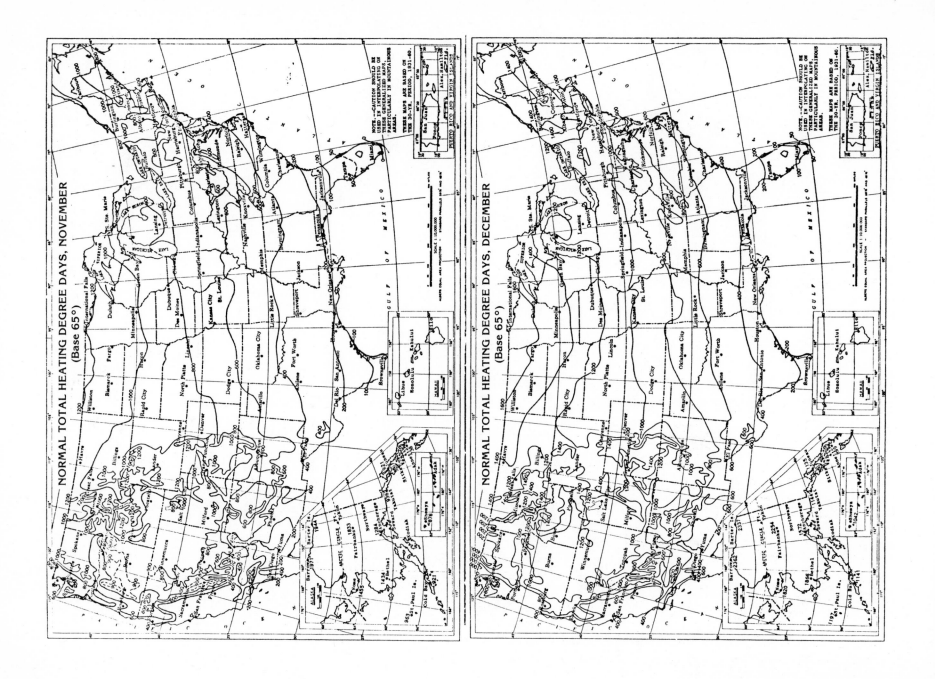

NORMAL TOTAL HEATING DEGREE DAYS, NOVEMBER
(Base 65°)

NORMAL TOTAL HEATING DEGREE DAYS, DECEMBER
(Base 65°)

APPENDIX C

CONCENTRATING COLLECTORS

The design of a flatplate solar energy collector is well within the capabilities of the architect; and, since it is probably going to quickly become the most proliferated application of the various solar energy technologies, then the architect should become well acquainted with it—if only in self-defense.

However, the concentrating collector is a more esoteric technology with a limited number of appropriate architectural applications. Yet it would be advantageous to have at least a passing acquaintance with some of its applications, limitations, and architectural implications.

With a flatplate collector system there is one, real physical limit—beyond the designed-in performance characteristics—that keeps the temperature relatively low. This physical limit is the fact that there is a finite maximum amount of sunshine that can be intercepted by each square foot of collector. For a flatplate collector, the amount of energy available is not a variable; it has a maximum intensity that cannot be exceeded. In order to achieve higher temperatures, the designers have to tinker with the amount of energy available; this is the crossover point from FPCs to concentrating collectors.

In order to raise the intensity level received by any one square meter of absorber, since the actual intensity cannot be changed, the energy received on a number of square meters has to be reflected to, or focused on, that one square meter of absorber. This concentrates the energy that is available over a large surface by focusing it on a smaller surface. This is why this type of device is called a concentrating or focusing collector (Figure C.1), and why the ratio of the area of intercepted sunlight to the area of the target or receiver is called the concentration ratio. For concentrators with tracking reflectors (Figure C.2), the area of intercepted sunlight is equal to the "mouth" area of the collector because they are always kept perpendicular to the incoming radiation.

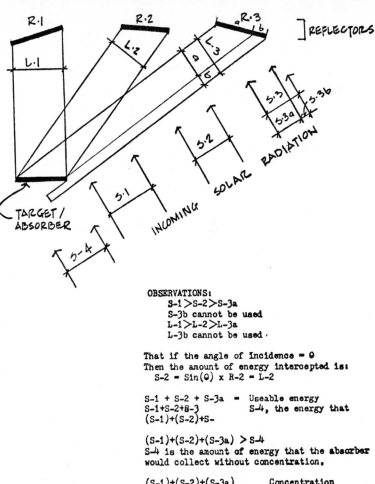

OBSERVATIONS:
$S-1 > S-2 > S-3a$
$S-3b$ cannot be used
$L-1 > L-2 > L-3a$
$L-3b$ cannot be used

That if the angle of incidence = 0
Then the amount of energy intercepted is:
$S-2 = Sin(\Theta) \times R-2 = L-2$

$S-1 + S-2 + S-3a$ = Useable energy
$S-1+S-2+S-3$ $S-4$, the energy that
$(S-1)+(S-2)+S-$

$(S-1)+(S-2)+(S-3a) > S-4$
$S-4$ is the amount of energy that the absorber would collect without concentration.

$$\frac{(S-1)+(S-2)+(S-3a)}{S-4} = \text{Concentration Ratio}$$

R-1 and R-2 have been sized so that the reflected light will completely cover the absorber, yet still not miss it. However, R-3(a+b) has the same surface area as R-1. This was done to illustrate that with a shallower angle of incident light, the reflector not only intercepts less sunlight ($S-1 > S-3(a+b)$), but that the shallow angle of reflected light also "dumps" part of the energy by overflowing the area of the absorber (L-3b). This is why R-2 is smaller than R-1.

Figure C.1

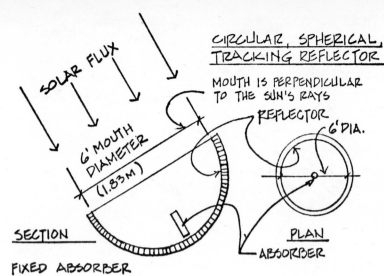

SOLAR FLUX

CIRCULAR, SPHERICAL, TRACKING REFLECTOR

MOUTH IS PERPENDICULAR TO THE SUN'S RAYS

REFLECTOR

6' MOUTH DIAMETER (1.83 M)

6' DIA.

SECTION

FIXED ABSORBER

PLAN

ABSORBER

REFLECTOR SURFACE AREA:
$\frac{1}{2}(4\pi R^2) = 56.5$ SQ.FT.
(5.25 M²)

"MOUTH" AREA = EFFECTIVE AREA OF INTERCEPTED SOLAR FLUX:
$\pi \frac{D^2}{4}$ OR $\pi R^2 = 28.3$ SQ.FT
(2.63 M²)

MOUTH AREA IS LESS THAN REFLECTOR AREA

SOLAR FLUX

REFLECTED LIGHT

MIRROR

NOTE:
FLAT REFLECTORS (MIRRORS) ARE USELESS IF THEY TRACK PERPENDICULAR TO THE SUN; THEY MERELY REFLECT THE LIGHT BACK TO THE SUN.

Figure C.2

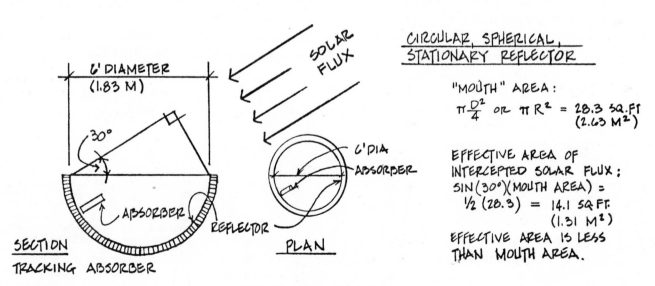

6' DIAMETER (1.83 M)

30°

SOLAR FLUX

CIRCULAR, SPHERICAL, STATIONARY REFLECTOR

"MOUTH" AREA:
$\pi \frac{D^2}{4}$ OR $\pi R^2 = 28.3$ SQ.FT
(2.63 M²)

EFFECTIVE AREA OF INTERCEPTED SOLAR FLUX:
SIN(30°)(MOUTH AREA) =
$\frac{1}{2}(28.3) = 14.1$ SQ.FT.
(1.31 M²)

EFFECTIVE AREA IS LESS THAN MOUTH AREA.

6' DIA
ABSORBER

ABSORBER

REFLECTOR

SECTION

TRACKING ABSORBER

PLAN

Figure C.3

For concentrators with tracking absorbers and stationary reflectors (Figure C.3), the area of intercepted sunlight varies; it is a function of the angle of solar incidence with the mouth of the collector and the mouth area. The concentration ratio is a direct indicator of how many times the energy available to the target has been increased. More energy means high possible temperatures. Some parabolic concentrating collectors have been able to generate temperatures above 3500°C (6300°F), but this is only with very high degrees of optical and mechanical precision. However, fairly crude approximations of parabolic collectors have been able to generate 500°C (930°F).[1]

Although a FPC cannot concentrate the sun's energy, it can utilize both the direct, incoming, parallel rays of sunlight and the diffuse light that is scattered on its way through the atmosphere. According to Liu and Jordan, approximately 40% of the solar energy available is of a diffuse nature.[2] A concentrating collector can only utilize approximately 60% of the available solar energy—the direct radiation portion of sunlight. This is because a concentrating collector requires parallel rays of light, and the only portion of solar insolation that is parallel is direct radiation.

Most concentrating collectors try to achieve some reasonable approximation of a parabolic reflector. This shape is chosen because the physics of the parabolic section creates a very sharp point of focus light that is parallel to the axis of the parabola (Figure C.4). This sharp point of focus means that only a small target is needed to absorb the concentrated energy. As the target gets smaller, the concentration ratio and the possible temperatures go higher. However, if the incoming light moves off axis (Figures C.5 and C.6), then the focus point begins to blur and disintegrate.

Since the sun moves throughout the day—in three dimensions (Figure C.7), the parabolic collector has to be constantly on the move to keep the light parallel to its axis of symmetry. This keeps the light converging at the focal point of the parabolic section. And, since this point is stationary relative to the reflector, it is the ideal point to place an absorber or target. It is at the absorber that the available energy that the reflector concentrates is taken up by the device to do work. The amount of energy available to do work depends on the area of the "mouth" of the collector or, at least, on the projection of the mouth area on a plane that is perpendicular to the sun's rays (Figures C.2 and C.3). However, the effectiveness of

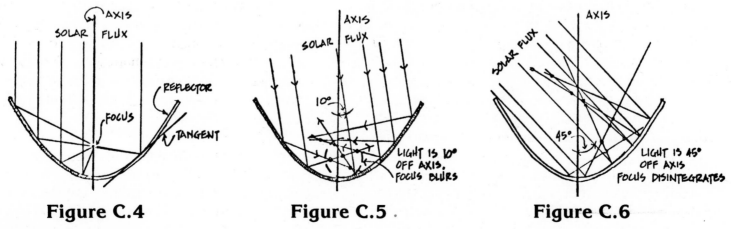

Figure C.4   **Figure C.5**   **Figure C.6**

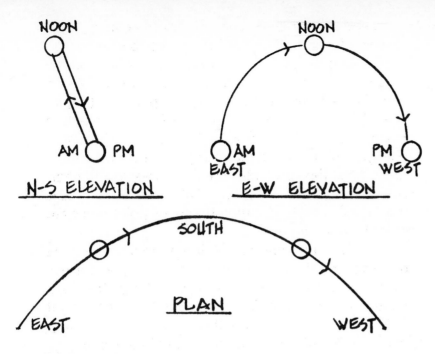

NOON

AM PM

N-S ELEVATION

NOON

AM PM
EAST WEST

E-W ELEVATION

SOUTH

EAST WEST

PLAN

Figure C.7

the collector in utilizing this energy depends on the accuracy with which the device is able to track the sun and on the optical quality of the reflector, because these factors affect the precision of the point of focus. The tolerances that are allowable in any particular collector system will be a function of the kinds of temperatures that have to be coaxed out of the device.

The higher the optical quality of the reflector and the more precisely the mechanism tracks the sun, then the sharper and smaller the precise point of focus becomes. If the point of focus can be kept small, then the target-absorber that intercepts this energy can be made correspondingly small. This means that the concentration ratios and the attainable temperatures become quite high.

The temperature that a collector is designed for would be the most important factor in establishing the concentration ratio for a collector. The solar fur-

nace at Odeillo and some very sophisticated dish collectors have been designed to produce temperatures in excess of 3000°C (5400°F). This is done with concentration ratios in the neighborhood of 1000. This means that the target is very small compared to the reflector. This would imply that either there is a lot of room available to spread out a huge tracking reflector, or banks of mirrors, if the absorber is rather large; or that there is a high quality, optical reflector that tracks the sun with great precision, giving a very sharp point of focus, if the absorber is small. Efforts to achieve high temperatures demand a very efficient absorber—and the smaller it is the better. The surface area of the absorber should be kept to a minimum in order to reduce inevitable radiative heat loss, which is a function of surface area. This level of sophistication is necessary for scientific research, but it is probably of no real consequence to the architect.

Although the architect will probably not be consulted on the design of the components for a concentrating collector, there are general comments that could give the architect a better informed judgement of the limits of concentrating collector systems.

For most applications, such high quality is not an overriding concern. All that is necessary is to determine that the quality is sufficiently high to meet the need for which it was designed. Imperfections in the shape of the mirror surface or in its finish do tend to reduce the sharpness of the point of focus. The precision of the tracking mechanism is also a major concern, yet it too can absorb a lot of imperfections before its performance is unacceptable for most applications.

The job of the tracking mechanism is to change the direction and the tilt of the concentrating reflector in order to keep the axis of the collector parallel to incoming direct radiation. One manifestation of the sophistication of the tracking mechanism is how often does it move to follow the sun. Does it do it continuously, moving smoothly, silently, so that the reflected image of the sun on the target never seems to move? Does it change once a minute with slightly perceptible bumps and jerks, jiggling the image on the target slightly? Or does it move once every fifteen minutes, or hourly, or even weekly, as is the case with parabolic troughs mounted E-W along their axis (Figure C.16)? How much should the image be allowed to move or become defocused (Figure C.5) before the collector is readjusted?

Economics will probably weigh heavily in any selection of concentrating collectors. In general, it is probably safe to assume that the larger the fabrication tolerances (aberrations) that can be allowed—while still delivering the temperatures needed—then the lower the final cost of the device will be. Even though this would yield collectors that might be technically inferior, even crude concentrators can develop temperatures of 500°C (930°F); and this is probably more than adequate for any remotely architectural applications.

One thing to keep in mind is that it would be incongruent—probably self-defeating—to mismatch levels of sophistication in the various components. Figure C.8 is a curve that compares cost to the optical quality of the mirroring surface of the reflector both in terms of the accuracy of the curved shapes generated and the quality of the applied mirroring material. Figure C.9 is a comparable curve for the comparison of the quality of the tracking mechanism to its costs. Both curves suggest that beyond a certain point, a real premium has to be paid for relatively small increments in quality. A mismatch would occur if a lot of money (Figure C.8, point B) were spent for a good quality reflector and it was put with a rather crude (Figure C.9, point A) tracking mechanism. The imprecision of the tracking assembly would probably negate the benefits that could have been reaped if the good reflector had been matched with an equally

Figure C.8

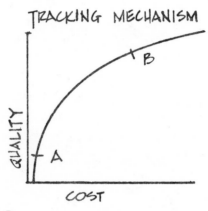

Figure C.9

good tracking mechanism. The "A" point on one curve should be paired with the "A" on the other; and the same applies to the "B" points on the curves.

An underlying assumption here is that there is little one can do with the target that will substantially effect the overall cost of the whole assembly. This is based on the fact that the sharpness of the point of focus is what determines the shape and size of the target-absorber, and this sharpness is a function of the quality of the reflector and the tracking mechanism.

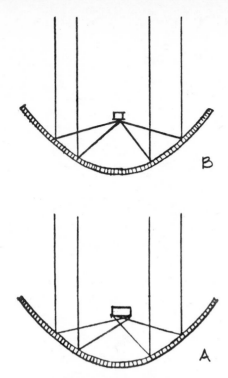

MOUTH AREA ON EACH
REFLECTOR IS: 5 M²
                    (53.8 FT²)
REFLECTOR "B" IS OF
HIGHER QUALITY THAN
REFLECTOR "A"

TARGET "A" HAS TO BE
LARGER THAN TARGET "B"
TARGET "A" = .1 M² (1.1 FT²)
TARGET "B" = .025 M² (.27 FT²)

CONCENTRATION RATIO
    B: 200
    A: 50

Figure C.10

The target has to be made large enough to intercept the whole focused image. If the image is large, then the target must be large; and this reduces the concentration ratio. If it is small then the target can be correspondingly small and the result will be high concentration ratios (Figure C.10). So, it is the quality of the mirrors/reflector and the tracking mechanism which determines the size of the focused image and the required target.

Another economic factor is just the sheer size of the devices. The concentration ratios for comparable collectors are directly related to the amount of materials needed for the reflector: by definition a concentration ratio of 200 requires twice as much reflector area as one with a concentration ratio of 100—if both have the same size target. The surface

area of a parabolic reflector almost doubles when the mouth area is doubled. With this added area the reflector would cost at least twice as much. In this way a comparison of concentration ratios, for the same quality reflectors, would give a quick estimate of their relative costs.

A paper presented at the ISES meeting in Fort Collins, Colorado in August, 1974, notes that, according to computer simulations they have been running on a spherical collector (as opposed to a parabolic), they have found that "efficiency always increases with concentration ratio although the marginal increase is quite small above a value of 50."[3]

This would suggest that a curve similar to those in Figures C.8 and C.9 could be drawn which would compare concentration ratios to efficiency (Figure C.11), which is a measure of how much energy is collected compared to the amount made available. This indicates that, as far as efficiency is concerned, there seems to be no significant advantage in generating high quality optics (since the optics set limits on the concentration ratios) to achieve marginal increases in efficiency, especially since it has the extreme disadvantage of high cost.

Figure C.10 illustrates the problem under discussion. Figure C.10A corresponds to the "A" points on

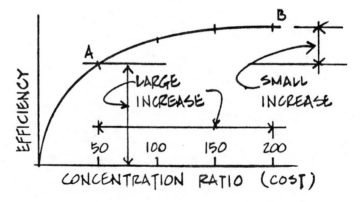

Figure C.11

the curves in Figures C.8, C.9, C.11. Figure C.10B corresponds to the "B" point on the curves in Figures C.8, C.9, C.11. It should be understood that Figure C.10B, although it would probably cost substantially more than Figure C.10A, cannot make any more energy available. They both reflect only 5 square meters of direct radiation towards a target. And, from Figure C.11 the extra little bit of efficiency shows that there is an insignificant increase in the amount of energy that is utilized, (Efficiency in Figure C.11 can also be read as a scale of the utilizable energy) since it tapers quickly after a concentration ratio of 50.

It is conceivable that the cost of producing a unit at point "B" could be twice as much as producing a similar unit at the quality of point "A". It would be a much wiser move to build **two** units at "A" quality than **one** at "B" quality; such a move would double the energy available and would probably far outstrip the one "B" unit in the production of utilizable energy. Although this hypothetical situation might be overdrawn, the point is that it is probably better to spend money to make more energy available at a lower temperature, than less energy available at a higher temperature; since the only real justification for having a high concentration ratio is to produce high temperatures.

From various experiments with solar parabolic concentrating collectors there is data that gives representative temperatures for different concentration ratios for circular parabolic collectors. One collector had a ratio of 174:1 (2.69m² mouth and .015m² target), and it produced an average temperature of 500°C (930°F).[4] Another collector had a ratio of 36:1 (1.1m² mouth with a .03m² target): it produced operating temperatures around 150°C (300°F). And with concentration ratios over 1000, temperatures have gone above 3000°C (5430°F). From this, it is reasonable to assume that temperatures well above 150°C can be obtained with a concentration ration of 50, and that it can be done with fairly good efficiencies and at a relatively reasonable cost (relative to the cost of other concentrating collectors). Again, these lower temperature collection devices should easily meet the criteria that any architect would be interested in.

So far, the discussion of concentrating collectors has been very generally concerned with those collectors which have a parabolic cross section; yet these

PARABOLIC "DISH"

AXIS
TARGET
REFLECTOR

CIRCLE
PARABOLA
CIRCULAR, PARABOLIC CONCENTRATOR

Figure C.12

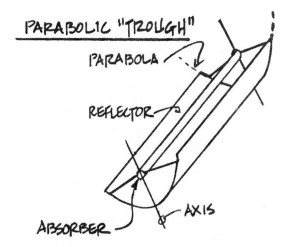

PARABOLIC "TROUGH"

PARABOLA
REFLECTOR
ABSORBER
AXIS

Figure C.13

generalities apply to all three basic parabolic types: the parabolic dish (Figure C.12), the parabolic cylinder or trough (Figure C.13), and a "fresnel" type reflector. However, there are some very important differences in terms of advantages of one over the other.

The parabolic dish (or circular parabolic) collector is probably the most difficult reflective surface to manufacture. The molds are difficult to machine accurately and it is difficult to apply mirroring materal uniformly over the two-way curved surface. In addition, the tracking mechanism has to be large enough to structurally support the reflector and the absorber and to stabilize the whole apparatus against vibration. The fact that the entire assembly must track requires a volume substantially larger than just the reflector (Figure C.14). But, in spite of these liabilities, it seems to be the most widely used concentrating collector. This is probably because most of the applications to date have been scientific in nature and they often require the high temperatures and sharp points of focus available with circular parabolic collectors. Another reason for their widespread use is the

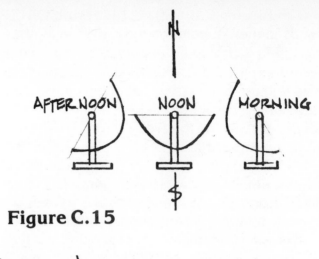

Figure C.15

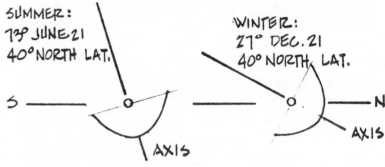

Figure C.16

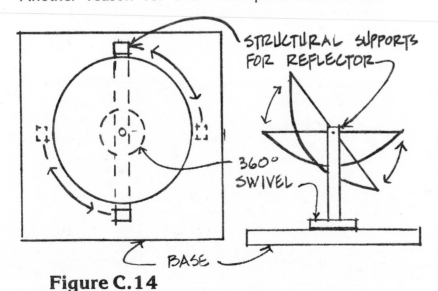

Figure C.14

availability of old army-surplus parabolic searchlights that are easily modified and operated in reverse as collectors. These old searchlights have precision mirrors and are relatively inexpensive—if they are still available.

The parabolic cylinder, or trough (Figure C.13), is much easier to fabricate than the circular, parabolic reflector. The surface is curved in only one direction, instead of the compound, two direction curvature of the parabolic dish. In addition, its linear format requires that it track in only one direction. It can be aligned North-South (Figure C.15) and it will turn to follow the sun throughout the day. Alternatively, the trough can be oriented East-West (Figure C.16) and

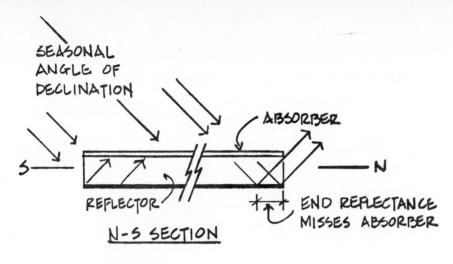

Figure C.17

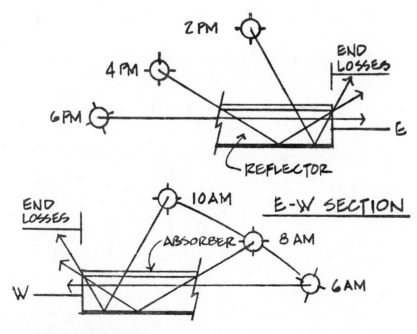

Figure C.18

it would not really need to track at all; it would just have to be adjusted occasionally (once a day, or once a week) to follow the seasonal changes in the declination angle of the sun.

The particular set up in Figure C.15 does not adjust itself to the changing declination angle of the sun. Because of this, it has to accept a small amount of end losses as a compromise (Figure C.17). These end losses are not significant if the collector is fairly long, since, the amount lost is very small if compared with the amount collected. The end losses are greatest for December 21, when the noon sun angle is at the lowest point for the year. The trough, however, has to adjust itself every half hour (Figure C.15) or so to keep the angle of incoming light as parallel as is reasonably possible to the axis of symmetry for the parabola (Figure C.4). Adjustments every half-hour keep the angle of deviation from parallel to 3.25°. Adjustments every hour allow twice as much deviation.

How much deviation should be allowed will have to be carefully studied in order to balance the quality of the reflector and the tracking mechanism with the size of the absorber. A ten degree angle (Figure C.5) shows severe defocusing; therefore 7.5° is probably a workable maximum. At this rate, hourly adjustment is required.

A cylindrical collector that is mounted East-West (Figure C.16) will generally have lower concentration ratios and lower temperatures for the same length. This occurs because of the compromises it makes. It has eliminated the tracking device by needing only occasional adjustment to the angle of declination of the sun, but the early morning and late afternoon sun is lost (Figure C.18) because of the low angles of incidence. These losses are similar to the end losses that occur in the case of the N-S oriented cylindrical collector. With only occasional adjustments to follow the declination of the sun, instead of having a diurnal tracking device, the reflector will produce a sharp point of focus only once or twice a day. At other times, the focus will be blurred as the altitude of the sun changes. If 7.5 degrees is an acceptable angle of deviation at the noon hour, the collector needs adjustment every two months (58 days). Monthly (28 days) adjustments will keep the angle of deviation to a maximum of 3.25°. The collector should be set so that the axis is parallel to 10 A.M. and 2 P.M. sun so that there are two daily maxima. Setting the collector for the noon sun gives only one daily maximum; the collector is in focus only once during the day. This daily process of focusing and defocusing means that temperatures and concentration ratios are generally lower than with a N-S oriented tracking trough.

Because of the blurring focus, the area of the target has to be designed to take advantage of most of the reflected energy. Figures C.19 and C.20 illustrate what happens as the focal point changes positions with respect to a stationary target (roughly equivalent to the broadening and blurring of an image as it would defocus [Figure C.5]).

Figure C.19 shows a target designed primarily for

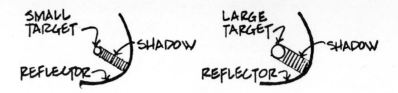

Figure C.21

a sharp point of focus. It works well for condition "A", but for conditions "B" and "C" there is a lot of energy slipping by that can never be utilized because the target was not designed to intercept it. Figure C.20, on the other hand, shows a target designed to intercept as much as possible of the reflected energy.

The idea is to make a target slightly larger, but only if it will mean a significant increase in available energy. There are drawbacks to making the target larger. A larger target throws a broader shadow on the reflector, blocking off energy (Figure C.21). A larger target also means that the concentration ratio, and consequently, the attainable temperatures, go down.

One advantage of the cylindrical concentrator,

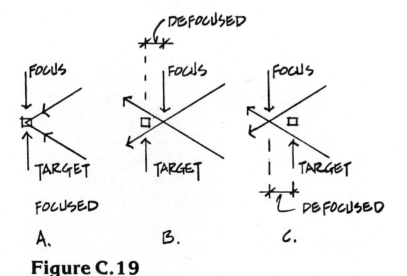

Figure C.19

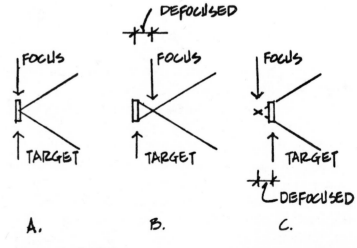

Figure C.20

when compared to the dish concentrator, is that the cylindrical collector can be readily added to if there is a need for additional energy. This has many implications for the architect. It means that solar installations do not have to be designed to meet expected future energy requirements if expansion is anticipated. When the time comes to expand, the simple addition of collector modules is all that is needed. This requires less in the way of capital equipment costs and would tend to lower initial cost of the first solar increment. This modularity also allows the designer the freedom to make a slight error in the beginning design if he is trying to avoid the cost penalty of oversizing the system. This freedom is not there in the design or specification of a dish collector. For example, if the energy output is lower than required by 5% with an arrangement of twenty cylindrical modules, then all that is probably necessary is the addition of one or two modules to raise the output of the system to the proper level. However, if he had depended on a single dish collector, anything less than the level of energy required would necessitate the addition of another dish. This would probably double the capacity of the system (and the costs) in order to pick up a missing 5%, since it would almost be absurd (Figure C.22) to design and build a collector to pick up just 5%. The dish system just does not lend itself very easily to incremental expansion. So, it forces an overdesign situation that probably would not be able to support future expansion anyway.

It should be noted that the expansion of cylindrical troughs makes more energy available, but it does not make any higher temperatures available (this is still a function of the concentration ratio). The stagnation temperature of the added segment will be the same as that of the initial components. Additional collector modules, however, will provide higher rates of output. The diagrams in Figure C.23 show how this could work. If the fluid flow rate through the absorber

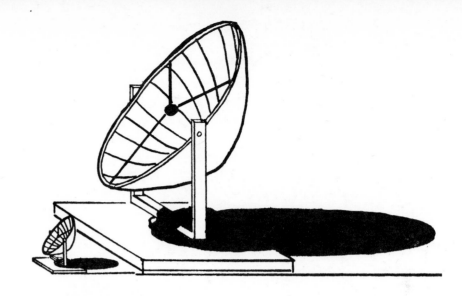

Figure C.22

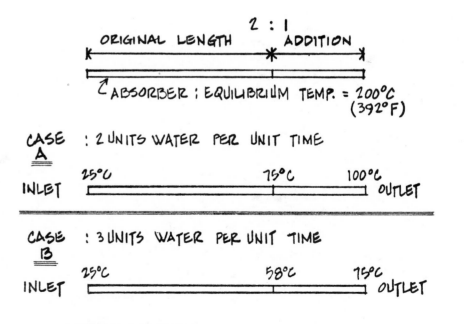

Figure C.23

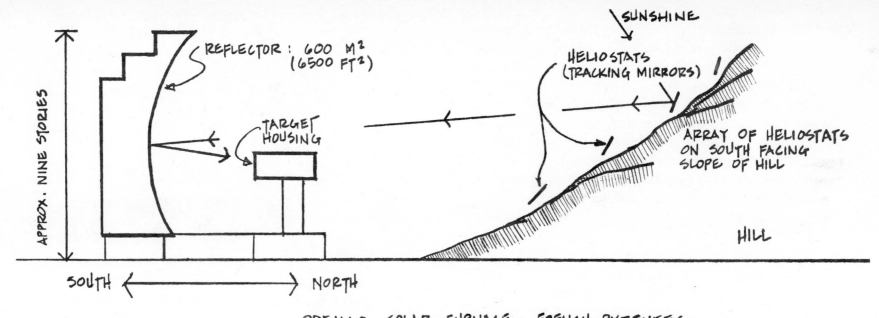

REFLECTOR : 600 M²
(6500 FT²)

SUNSHINE

HELIOSTATS
(TRACKING MIRRORS)

TARGET
HOUSING

ARRAY OF HELIOSTATS
ON SOUTH FACING
SLOPE OF HILL

APPROX. NINE STORIES

HILL

SOUTH ⟷ NORTH

ODEILLO SOLAR FURNACE · FRENCH PYRENEES

Figure C.24

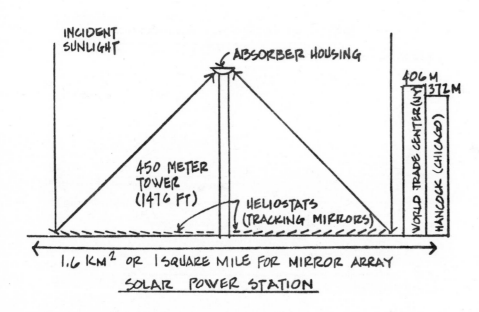

INCIDENT
SUNLIGHT

ABSORBER HOUSING

406 M
372 M

WORLD TRADE CENTER (NY)

HANCOCK (CHICAGO)

450 METER
TOWER
(1476 FT)

HELIOSTATS
(TRACKING MIRRORS)

1.6 KM² OR 1 SQUARE MILE FOR MIRROR ARRAY

SOLAR POWER STATION

Figure C.25

were to stay the same (Figure C.23A), then the outlet temperature after the length has been increased will be higher—as long as the operating temperature of the absorber is significantly higher than the fluid temperature to facilitate efficient heat transfer. If the fluid flow rate were increased (Figure C.23B), then it would be possible to maintain the same outlet temperature as the initial array with more working fluid available at the end of this heat collection process. (Note: the numbers in Figure C.23 are for illustrative purposes only.)

The "fresnel" type of reflector, as it is referred to here, has as its only approximate examples large solar furnace installations such as Odeillo (Figure C.24) in the French Pyrenees and central power stations (Figure C.25) that have been proposed for the arid Southwest United States.

The fresnel concept of segmenting and compressing the profile of a lens can be transferred to the

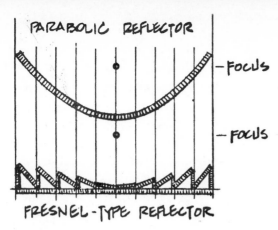

PARABOLIC REFLECTOR

—FOCUS

—FOCUS

FRESNEL-TYPE REFLECTOR

Figure C.26

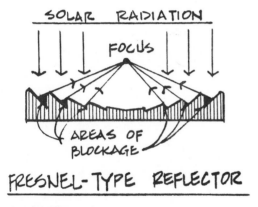

SOLAR RADIATION

FOCUS

AREAS OF BLOCKAGE

FRESNEL-TYPE REFLECTOR

Figure C.27

parabolic reflector concept (Figure C.26) to yield a fresnel-type reflector. If plates could be stamped or extruded and coated with a highly reflective surface, then this might be an economical approach to a concentrating collector. The main drawback to such a collector is that there would be some blockage of light being reflected to the focal point (Figure C.27). To what extent this would be a problem remains to be seen, it would probably require some experiments to determine the actual effect of the blockage on performance.

However, there does not seem to be any literature on such a collector, so it must be assumed that such a device has not as yet undergone experimental testing. It seems, though, that its limitations would be the same as a circular parabolic collector, (or the cylindrical troughs, if the patterns were etched in a linear manner). One important factor that might affect the performance of such a device would be the reflectivity of the surface; it could be substantially lower than a mirror coated surface. With substantially lower fabrication costs, such problems might still prove to be an acceptable compromise between economics and energy output.

A fresnel type application that receives much attention now, is the large fields of mirrors with central towers. In these schemes, large arrays of flat, tracking mirrors are deployed over many square kilometers. These mirrors track the sun in the same way that a point on the surface of a parabolic reflector would (Figure C.28): if the incoming light is parallel to the axis, then the tangent at any point (Figure C.29) —reflecting the light at an angle equal to the incidence angle— will redirect the light to the point of focus. The scale of the field of reflectors is so large that each rather large tracking mirror ($5m^2$) can be considered not only a small point on a fresnel reflector but also as the precise point of tangency— so that it is flat in theory as well as in fact. This array is readily seen as a large dispersed fresnel reflector (Figure C.25) with its focus point at the top of a very large tower. The purpose of such large scale solar energy systems is to generate electrical power on a very large scale. There are many problems yet to be worked out in such a scheme, but at the current time, there are plans to build a pilot project. It should provide answers for some of these problems while rigorously testing the workability of a solar central power generating station.

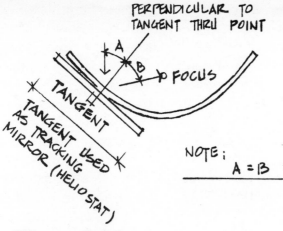

PERPENDICULAR TO TANGENT THRU POINT

TANGENT

TANGENT USED AS TRACKING MIRROR (HELIOSTAT)

FOCUS

NOTE:

A = B

Figure C.28

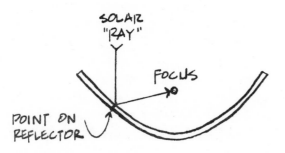

SOLAR "RAY"

FOCUS

POINT ON REFLECTOR

Figure C.29

One of the basic concerns about such a concept is the fact that a great deal of time, energy and raw material has gone into an effort to concentrate a naturally dispersed energy source. This energy is then redistributed over an utility network so that the energy, again, is dispersed. The whole process has been estimated, variously, to be between 10 and 15 per cent efficient (Figure C.30) —from the square foot of incoming sunlight to that piece of hot buttered toast at breakfast.

It is astonishing that such measures are necessary when less power would be lost if this same amount of

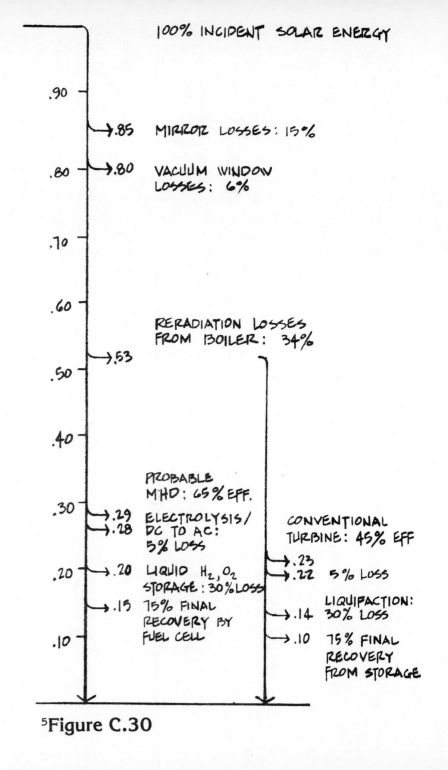

100% INCIDENT SOLAR ENERGY

.90

.85 MIRROR LOSSES: 15%

.80 VACUUM WINDOW LOSSES: 6%

.70

.60

RERADIATION LOSSES FROM BOILER: 34%

.53

.50

.40

PROBABLE MHD: 65% EFF.

.30 → .29 ELECTROLYSIS/ DC TO AC: 5% LOSS
→ .28

CONVENTIONAL TURBINE: 45% EFF

→ .23
→ .22 5% LOSS

.20 → .20 LIQUID H_2, O_2 STORAGE: 30% LOSS

LIQUIFACTION: 30% LOSS
→ .14

→ .15 75% FINAL RECOVERY BY FUEL CELL

.10

→ .10 75% FINAL RECOVERY FROM STORAGE

[5]**Figure C.30**

dispersed solar energy could be used at the very sites on which it falls. To do this, however, would require that each site installation at least have someone who would know how to run and operate the system. This might be acceptable for commercial or industrial applications, but it would be almost too much to expect that the average American household would be able, or even willing, to take the responsibility for running its own power or heating and cooling systems. The fact remains that the average American wants his power on tap at just the flip of a switch. It is also a fact that, unless solar power can be converted into the readily usable form of electrical energy, solar power will not be able to make significant inroads into the energy demands of major industrial countries.

NOTES

1. F. Daniels, **Direct Use of the Sun's Energy** (New York: Ballantine Books, 1975), p. 42.

2. B. Y. H. Liu and R. C. Jordan, "Long-Term Average Performance of Flat-Plate Solar Energy Collectors," **Solar Energy** (Vol. 7, No. 2, 1963), p. 53.

3. **Technical Program and Abstracts** (Fort Collins, Colorado: ISES Annual Meeting, August, 1974), p. 24. and personal notes on a paper by J. F. Kreider and G. Steward "The Stationary Reflector/Tracking Absorber Solar Concentrator."

4. F. Daniels, **op. cit.**, p. 46.

5. Figure was adapted from A. F. Hildebrant, et. al., "Large Scale Concentration and Conversion of Solar Energy," **Transactions of the American Geophysical Union** (Vol. 53, No. 7, July, 1972), p. 691.

APPENDIX D

DESIGN OF A SOLAR HOUSE

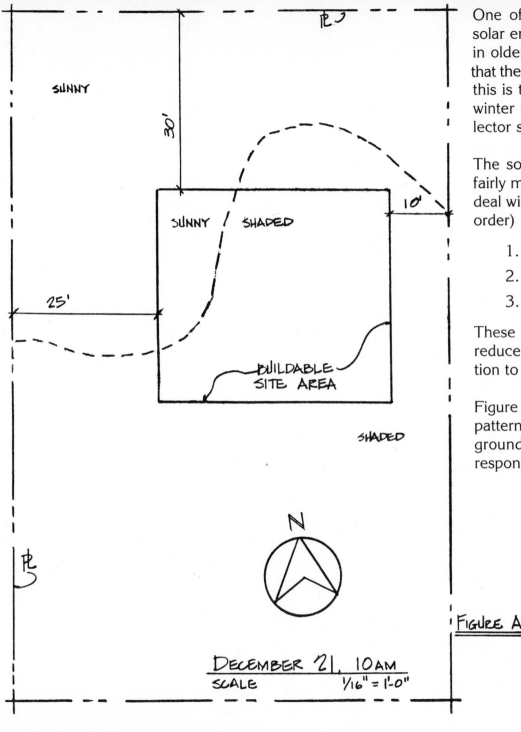

SUNNY

30'

SUNNY / SHADED

10'

25'

BUILDABLE
SITE AREA

SHADED

N

DECEMBER 21, 10 AM
SCALE   1/16" = 1'-0"

One of the obvious difficulties involved in utilizing solar energy for residential space heating, especially in older, well established neighborhoods, is the fact that the trees in the area may have reached maturity. If this is the case, one should investigate the effect of winter shading on the site at the level of one's collector surface.

The south and east sides of the site are bound by fairly mature trees—42-48 feet in height. The way to deal with the tree problem is three-fold: (in preferred order)

1. Move the building
2. Top the trees
3. Cut them down/Abandon the solar scheme

These trees could possibly block or substantially reduce chances that there could be sufficient insolation to warrant collection.

Figure A shows the December 21st 10 A.M. shade pattern as it would appear on a plane 20' above the ground—a probable height for collectors. In response to alternative number one this means that

FIGURE A

the building should be in the northwest corner of the site area; but, once the setbacks are determined, choices in moving the building become severely limited.

In response to alternatives 2 & 3, the trees on the east present the most serious problem for collection, but they are on the adjacent parcel of property and cannot be either topped or removed. Since the east tree pattern cannot be altered, and since the trees on the south do not present too severe a liability to collection, the decision was made to leave the trees alone—but monitoring the growth of the south side trees with an eye toward keeping them at their present height.

At first glance the situation looked bleak, but continued examination uncovered some heartening facts:

1. All the trees are deciduous, only the branches and trunk remain to obscure sunlight.

2. December 21st is the worst day of the year, all things being equal, for trying to collect sunshine.

3. The average possibility of sunshine in December is 43%; thus the chances are better than half that the sun will not even shine that day.

4. Things can only get better.

Ivy trellises can be used as barriers to summer heat on the west wall. The leaves follow the sun, shading the wall, and their fabric is loosely woven allowing air to circulate freely between the house wall and the trellis. During the winter, the ivy will die and allow full penetration of the sun to the wall, warming the house.

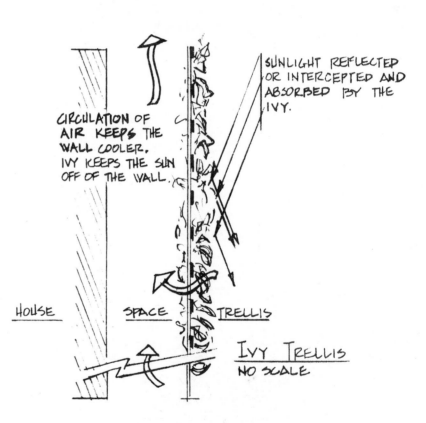

SUNLIGHT REFLECTED OR INTERCEPTED AND ABSORBED BY THE IVY.

CIRCULATION OF AIR KEEPS THE WALL COOLER. IVY KEEPS THE SUN OFF OF THE WALL.

HOUSE   SPACE   TRELLIS

IVY TRELLIS
NO SCALE

One of the generating parameters along with the site, for this design, was the fact that it was to be approached by following F.H.A. guidelines as a minimum. In the FHA-MPS for one and two living units, one of the purposes is to assure that, "other factors, such as the appropriateness of the dwelling to the site and to the neighborhood and the anticipated market acceptance of the property as a whole are considered in F.H.A. underwriting analysis." In addition, the Maryland Terrace Subdivision (of which this lot is a part) has the power to prevent construction if they deem the building inappropriate to the neighborhood.

With the neighborhood in mind, an effort was made to conform as closely as possible to the traditional housing forms in the area, deviating only when necessary. This means that a 40° roof angle would be deemed more acceptable than 50° or 60°, the house should be two stories and there should be roughly 1100 square feet per floor in order to compare with the house next door. If one constrained himself to one of the more traditional forms, this approach might also demonstrate the feasibility of retrofit solar installations. In addition, a forty degree roof angle would be better during the summer months—for the anticipated expansion of the system to include power needs for air conditioning.

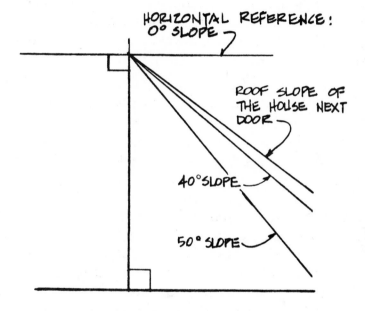

From solar studies made by Lof and Tybout, there is a working hypothesis that if 60-70% of the load for a heating season can be carried by the solar heating system, then one probably has a system that is economically favorable when compared with electric heat and one that is competitive with gas and oil, depending on regional availability.

With the rationale that the house next door could be considered "traditional," this designer began a solar feasibility study for space heating using the house next door as a basis.

This unit has two full floors and a heated basement at 1125 square feet per floor. National Electric Manufacturers Association (NEMA) recommended levels of building insulation were equaled or bettered as a basis for calculating heat losses. Ten percent of the floor area of the two upper floors and 2% of the basement area were used to calculate the required insulated glass and one air change an hour was assumed. (See Table I, D-5) These calculations showed that a structure of this kind would lose approximately 17,500 BTU/Degree Day. With this estimate at hand it was determined that 800 square feet of collector surface area could collect enough heat (through a total system efficiency of 33%) to accommodate 100% of the heat loss for February, 80% for January and December, and 115% for November. Thus, 800 square feet was deemed more than acceptable, and the problem became one of developing a design that would architecturally integrate this collector surface area in a fashion acceptable to the neighborhood.

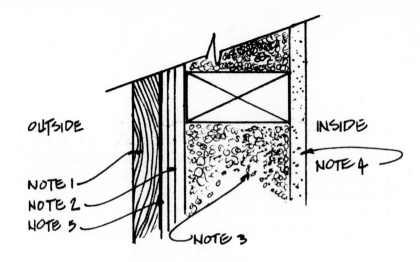

RESISTANCE TO HEAT TRANSMISSION:

	OUTSIDE AIR	.17
1	3/4"×10" LAPPED SIDING	1.05
2	25/32" INSULATION BOARD	2.06
3	APPROX. 3" FOAMED INSULATION	24.00
4	1/2" GYPSUM BRD.	.45
5	TAR PAPER	.06
	INSIDE AIR	.68

TOTAL RESISTANCE = 28.47 = R_T

$1/R_T = U_{FACTOR} = .035$ BTUH/FT².°F HEAT LOSS

WALL DETAIL · THERMAL PROPERTIES

NO SCALE

ELEMENT	FT^{1,2,3}	U FACTOR NEMA	U FACTOR DESIGN	BTUH/°F	BTU/DD	BTU/70DD	COMMENTS
EXTERIOR WALL	140'x 22' (3080²)	.07	.035	107.8	2587	181,104	2' IS BSM'T ABOVE GRADE
CEILING	1125²	.05	.035	39.4	945	33,075	35°F = MAX ΔT
BSM'T WALL B.GRD	840²	4 BTUH/FT²		3360	80640	80640	STEADY STATE
FLOOR	1125²	2 BTUH/FT²		2250	54000	54000	STEADY STATE
GLASS 2% BSM'T	22.5²	.56		12.6	302	21168	STORM
10% HOUSE	225²	.56		126	3020	211680	STORM
AIR CHANGE	21093³	.018		379	9112	637875	¾ VOLUME
TOTALS					20,606	1,219,542	

FOR ESTIMATING PURPOSES ONLY: 1,219,542/70 = 17,422 BTU/DD

NOTE: NEMA – U-FACTORS RECOMMENDED BY THE NATIONAL ELECTRICAL MANUFACTURERS ASSOCIATION.

DESIGN – DESIGNATES PROJECTED U-FACTORS FOR THE CONSTRUCTION USED IN THE SOLAR DESIGN

Before the actual design process could begin it was necessary to establish what the climate of a site in University City, Mo. would demand from, as well as supply to, a solar heating system. It also meant that system parameters for the collection and utilization of solar energy had to be formulated. So that this might be done, it was necessary to construct the weather data for urban St. Louis. Much of this information came from a Columbia weather station and had to be corrected for peculiarities in the urban climate.

The questions to be answered are: how many BTUs can one square foot of collector deliver to actual house use, and how will it vary?

In order to do this one needs to know:

DEGREE DAYS

INSOLATION

COLLECTOR TILT

SYSTEM EFFICIENCY

STORAGE CAPABILITIES

ST. LOUIS

MONTHLY TOTAL
930.9
789.9
648.8
300.3
112.3
27.3
—
—
54.6
225.6
573.3
846.3

← .91

LAMBERT

MONTHLY TOTAL	MEAN DEG. DAYS	MONTH
1023.	33	JAN.
868	31	FEB.
713	23	MAR.
330	11	APR.
124	4	MAY
30	1	JUNE
—	—	JULY
—	—	AUG
60	2	SEPT
248	8	OCT
630	21	NOV
930	30	DEC.

In the Solar Design Handbook, there is a chart that records the Mean Degree Days by months for Lambert Airport in St. Louis. However, it is a well-known fact that cities are often warmer than their hinterlands. In the case of St. Louis, the airport normally experiences 4900 DD per year while the city has only 4500.

To arrive at the degrees for each month in urban St. Louis, one has to multiply the monthly mean for Lambert by the number of days in the month and get a monthly total of degree days for the airport. These monthly totals of DD are then converted to monthly totals of Degree Days for St. Louis by multiplying the Lambert totals for Months by the ratio of DD in St. Louis to the DD at Lambert: 4500/4900 = .91.

ST. LOUIS

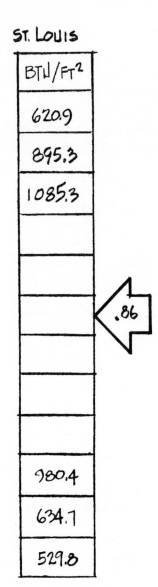

BTU/FT²
620.9
895.3
1085.3
980.4
634.7
529.8

.86

COLUMBIA

BTU/FT²	MONTH
722	JAN
1041	FEB
1262	MAR
1738	APR
2071	MAY
2308	JUNE
2293	JULY
1904	AUG
1461	SEPT
1140	OCT
738	NOV
616	DEC

From the weather station in Columbia, Mo. there is available the average daily number of BTUs incident on a horizontal surface—for each month. However, because of differences in climate between Columbia and St. Louis, it was necessary to gather at least some insolation data for St. Louis so that a correlation might be established. The scant information available for St. Louis was compared to Columbia and it seems that St. Louis generally receives approximately 86% of the level of insolation that is available in Columbia, Mo. So to arrive at the correct St. Louis average daily number of BTUs on a horizontal surface, one has to multiply the Columbia averages by .86.

40° TILT

BTU/FT²	MULTI. FACTOR
1415	2.28
1611	1.80
1476	1.36
1588	1.62
1320	2.08
1314	2.48

HORIZONTAL

BTU/FT²	MONTH
620.9	JAN
895.3	FEB
1085.3	MAR
980.4	OCT
634.7	NOV
529.8	DEC

ST LOUIS

Once one has the average daily number of BTUs available on a horizontal surface in St. Louis, then there is the task of optimizing solar energy collection. One should realize that the angle of the collector plate (See Figure B1 and B2) to the path of the sun's rays can substantially effect the level of energy one can collect. A rule of thumb for collector tilt for wintertime heating is: add 10° to the latitude of one's position. For St. Louis, with a latitude of 39.5° the rule of thumb states that collector tilt would be about 50° with respect to the ground. However, the roof profile of a 50° angle seems visually inappropriate to the neighborhood; so this designer used a 40° tilt and corresponding multiplying factors to correct the average daily BTUs on a horizontal surface to the average daily BTUs on a 40° surface.

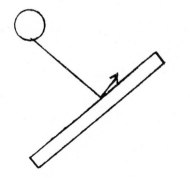

B-1
"TILTED"

B-2
"HORIZONTAL"

A PLANE TILTED MORE PERPENDICULARLY TO THE SUN'S RAYS WILL ABSORB MORE ENERGY

DELIVERED

BTU/FT²
466.9
531.6
487.0
524.0
435.6
433.6

INTERCEPTED

BTU/FT²	MONTH
1415	JAN
1611	FEB
1476	MAR
1588	OCT
1320	NOV
1314	DEC

.33

NOTE: IF STORAGE IS IN THE DWELLING, THE
THE SYSTEM'S HEAT LOSS (THE .66) IS
RELEASED TO THE DWELLING — IMPROVING
EFFICIENCY

After one has determined how many BTUs are being intercepted by his collector, then he has to determine what sort of efficiency his overall system is capable of delivering. The nature of thermal energy is such that it tries to dissipate itself to its surroundings. So, although one might intercept 3 BTUs at the collector surface, only 1 BTU probably arrives at storage to be accumulated and saved. This ratio 3:1 or 33% is about right, too. One can nominally design the collector so that about one-half of what it receives it is able to pass through to the storage system, but some of this thermal energy is lost in transit so that approximately 66% of the power leaving the collector probably reaches the house through storage and is considered usable. $(.66) \times (.5) = .33$. This shows that the collector will be able to deliver ⅓ of the BTUs it intercepts to the house and that this will vary with respect to insolation levels, latitude, and collector tilt.

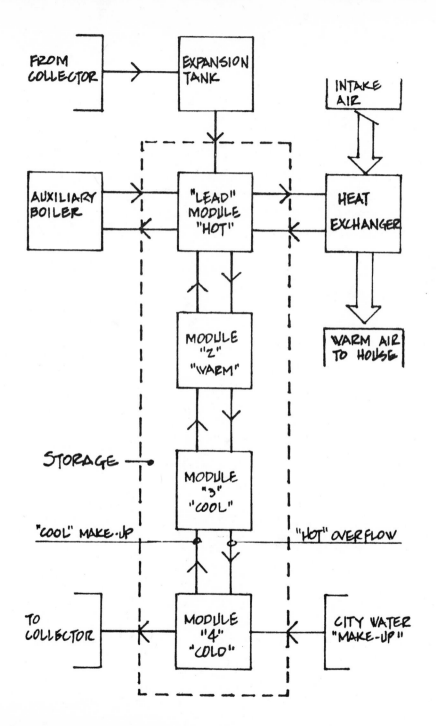

The solar heating system in this design utilizes water as both a collection and storage medium. This system was chosen primarily for the fact that water is relatively inexpensive and using it would require only one heat exchanger—from water to the forced air heat distribution system.

The system is designed to begin collection when the absorber plates reach a temperature 10°F higher than that of the storage temperature and to shut off and drain the collectors if the temperature drops below this 10° margin.

The volume of water that could be stored is approximately 150 cubic feet. At maximum fill, the storage unit would hold 9,375 pounds of water and it would be distributed over 42.25 square feet (6.5' × 6.5'). Anticipating that the water may become heated to 200°F and recognizing that the water has to be maintained at 85°F for heating purposes, there exists the potential storage capacity of 1,078,125 BTU.

This stage of the design is accomplished by having determined:

1. The heat loss of the structure in BTUs/Degree Day.

2. Total monthly number of Degree Days for University City.

3. The percentage of the heating load that should be carried by solar power.

4. The number of average daily deliverable BTUs to storage.

5. The critical period of the year where the difference between supply and demand is greatest.

These numbers will be most important in establishing one's collector area. Other factors which make precise sizing difficult are:

1. Probability of sequential cloudy days.

2. Changing economics of fossil fuel supply, demand.

3. Changing economics of collector construction.

The house relies on its neighbor's trees for shading the east side during the summer mornings. On the west side, there is a 20 foot high ivy trellis (above 20' there is only attic). This ivy is seasonal. It will die back during the winter, allowing for maximum sun on the wall; and it will grow back during the spring and provide full coverage by summer.

Since south is the prevailing summer wind direction, the larger fenestrations were placed on the south wall for maximum ventilation. The windows on the other three sides are noticeably smaller, but it is expected that the movement of summer winds along the sides and over the top will generate a slight negative pressure to help pull the winds in through the south wall. Since the prevailing winter wind is northwest, the windows on the north and west are kept small to reduce infiltration from winter wind pressure.

The large overhangs on the south are designed to shade the windows during the summer (see elevations) and to admit as much sun as possible during the winter. The summer (June 21) sun angle at noon is 73°; during the winter (December 21) the sun angle is 27°. In addition, the relatively high wall placement of most windows allows for deeper penetration of natural light into the interior.

In order to help buffer noise from the outside, especially from the freeway, the dining room, kitchen and laundry—which generate their own background noise when in use—were placed on the north wall. It is assumed that their noise levels would pleasantly mask any intruding freeway noise. The bedrooms on the north wall will be used primarily at night when the freeway noise level should be of little or no consequence. The living room, study and den are conver-sation/quiet areas and are to the south, where they can benefit from the buffering effects of the north-side rooms. Spaces on the north side also tend to be warmer due to their use. This fact could help balance the heavier heat losses expected from a north, windward wall. If insulating shutters were used at night to close the large south windows, then there could be a positive heat gain during the day from the incoming sunshine. In addition, for more effective energy utilization, the heating and cooling systems are zoned for the major use areas—rather than constantly conditioning the entire volume of the house.

The suggested lighting system would be almost entirely fluorescent and would provide overall circulation lighting of 15 foot candles/square foot. It would provide for task lighting at 70 foot candles at selected areas that would total 250 square feet (i.e. this is 250 square feet of actual working plane illumination). This will be achieved through counter mounted lights and table lamps. A few, portable lamps would be the only incandescent light sources.

SOLAR HOUSE HEAT LOSS CALCULATIONS

ELEMENT	FT^{1,2,3}	U FACTOR	BTUH/°F	BTU/DD	BTU/70DD	COMMENTS
EXTERIOR WALL	145' x 18.5' 2682^2	.035	93.8	2251	157,700	
CEILING	1045^2	.035	36.6	878	30,723	35°F = MAX ΔT
SLAB EDGE	145'	—	51.7	1242	.87000	25 BTUH/FT ΔT = 70°F
GLASS STORM INSUL	140.75^2 55.25^2	.56 .65	78.8 35.9	1891 862	132384 60334	WINDOWS ¼" DBL. DOORS
AIR CHANGE	14,832^3	.018	266.9	6406	448490	¾ VOLUME
SUBTOTALS			563.7	13,530	916,631	

ELEMENT	FT 1,2,3,	U FACTOR	EQUIV. T.	HTM*	BTUH	COMMENTS
SOUTH WALL	468²	.035	23.6 ΔT	.83	388	SHADED BY OVERHANGS
SOUTH GLASS	70.5²	.61		9	634	
EAST WALL	608²	.035		.83	505	SHADED BY NEIGHBOR'S TREES
EAST GLASS	26.5²	.61		10	265	
NORTH WALL	611²	.035		.83	509	
NORTH GLASS	48²	.61		8	384	
WEST WALL	605²	.035		.83	502	SHADED BY TRELLIS
WEST GLASS	30²	.61	↓	10	300	
CEILING	1045²	.035	44.0	1.54	1610	VENTED ATTIC
AIR CHANGE	9888³	.018	20		3333	½ AIR CHANGE/HR.
2-3068 DOORS	40²	.43	23.6	10.1	404	
PEOPLE	2 PERSONS/BDR @ 4 BDR = 8 PEOPLE @ 400 BTU'S PER PERSON				3200	FROM McG & STEIN
KITCHEN					1200	FROM McG & STEIN
					13234	SUBTOTAL

LIGHTING	SQ FT	TYPE	FOOT CANDLES	WATTS/ SQ FT	WATTS	.3 USE FACTOR	BTUH	COMMENTS
CIRCULATION	1964²	DIRECT/IND.	15	1.1	2160	648	2203.2	.3 FACTOR IS DESIGNER'S JUDGEMENT OF USE AVERAGE OVER 24 HRS.
TASK	200²	DIRECT/IND.	70	4.9	980	342	1162.8	

SENSIBLE HEAT GAIN   16,600 BTUH
LATENT HEAT GAIN (@.3) + 4980 BTUH
                     21580

CAPACITY MULTIPLIER   .75
                16185 BTUH = REQ'D COOLING

*HTM IS COMPUTED BY MULTIPLYING U FACTOR X EQUIVALENT TEMPERATURE. SEE P173 McG & S. GLASS HTM FROM TABLES.
NOTE: WASHER & DRYER SPACE VENTED TO OUTSIDE

SOLAR HOUSE HEAT GAIN CALCULATIONS
SUMMER AIR CONDITIONING LOADS

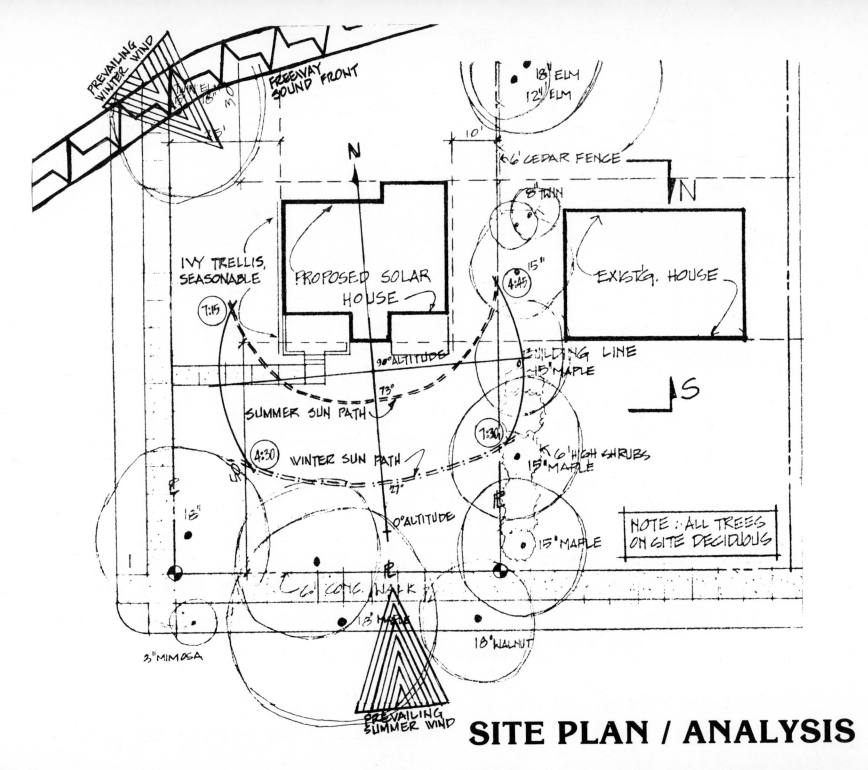

PREVAILING
WINTER WIND

TWIN ELM
18" 30'
15'

FREEWAY
SOUND FRONT

18" ELM
12" ELM

N

10'

6' CEDAR FENCE

N

8" TWIN

IVY TRELLIS,
SEASONABLE

PROPOSED SOLAR
HOUSE

4:45   15"

EXIST'G. HOUSE

7:15

90° ALTITUDE

BUILDING LINE
15" MAPLE

S

73°

SUMMER SUN PATH

7:30

4:30   WINTER SUN PATH

6' HIGH SHRUBS
15" MAPLE

27°

18"

0° ALTITUDE

15" MAPLE

NOTE: ALL TREES
ON SITE DECIDUOUS

6" CONC. WALK

18" MAPLE

18" WALNUT

3" MIMOSA

PREVAILING
SUMMER WIND

SITE PLAN / ANALYSIS

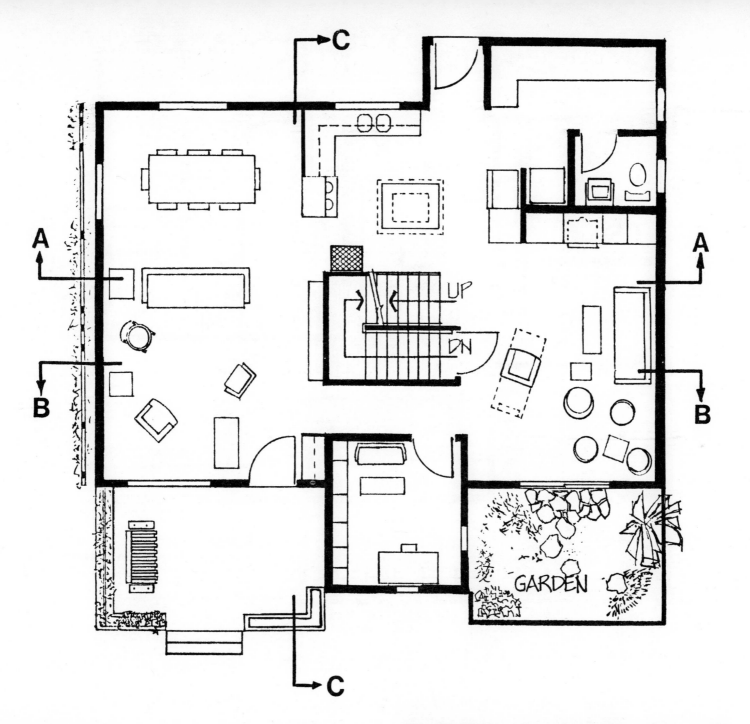

FIRST FLOOR PLAN

207

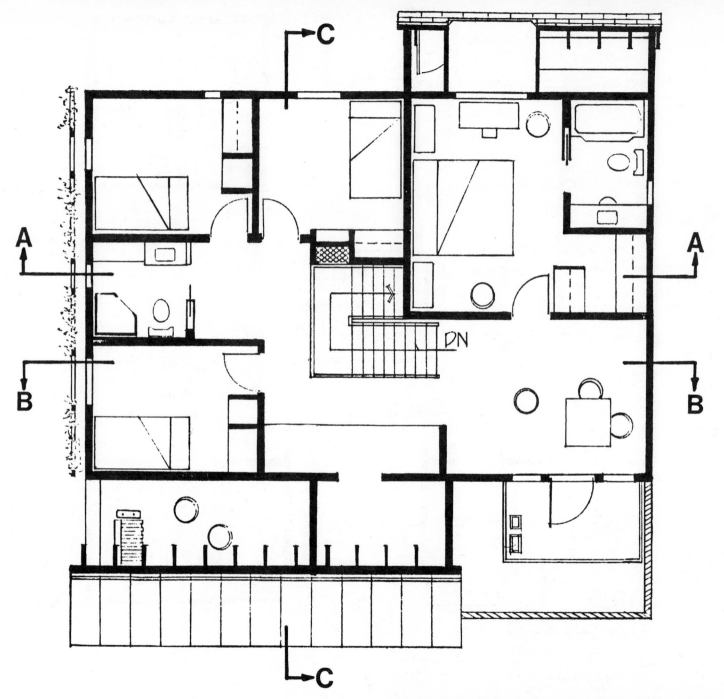

SECOND FLOOR PLAN

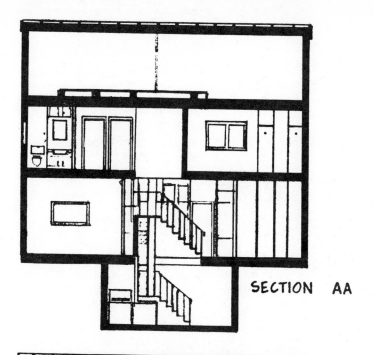

SECTION AA

SECTION CC

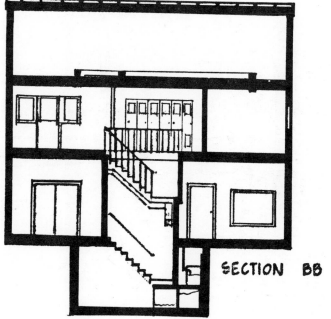

SECTION BB

SECTIONS

WEST SECTION - ELEVATION

A: Solar Noon Angle
   December 21
B: Solar Noon Angle
   June 21

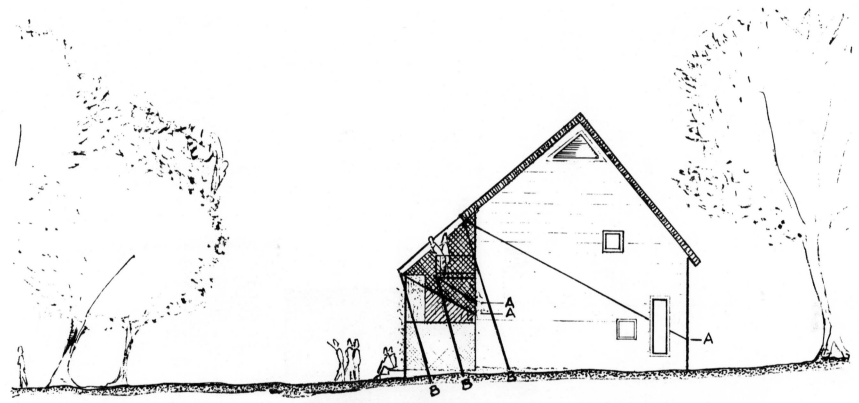

EAST SECTION - ELEVATION

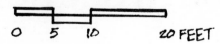

0   5   10   20 FEET

A: Solar Noon Angle
   December 21
B: Solar Noon Angle
   June 21

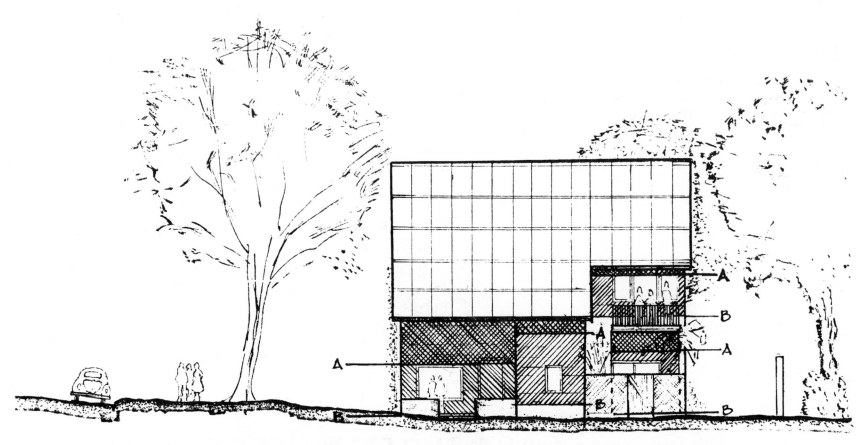

SOUTH SECTION - ELEVATION

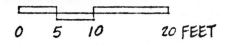

0   5   10   20 FEET

A: Solar Noon Angle
   December 21
B: Solar Noon Angle
   june 21

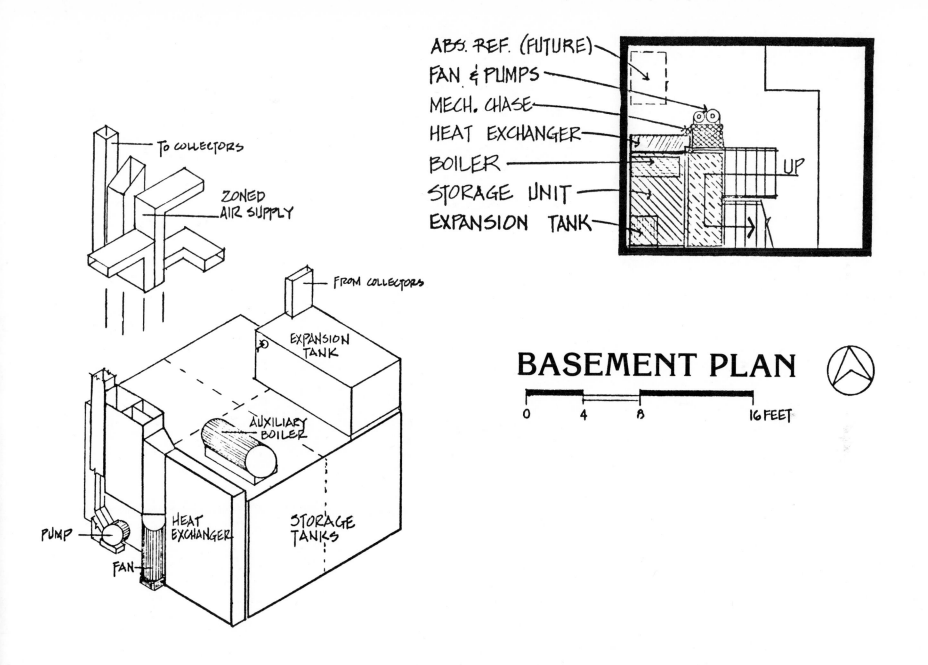

To COLLECTORS

ZONED AIR SUPPLY

FROM COLLECTORS

EXPANSION TANK

AUXILIARY BOILER

PUMP

FAN

HEAT EXCHANGER

STORAGE TANKS

ABS. REF. (FUTURE)
FAN & PUMPS
MECH. CHASE
HEAT EXCHANGER
BOILER
STORAGE UNIT
EXPANSION TANK

UP

BASEMENT PLAN

0   4   8   16 FEET

MECHANICAL SYSTEM

APPENDIX E

ENERGY CONSERVATION
ANALYSIS TECHNIQUE

TYPICAL HOUSE < 5 YEARS OLD

ASSUMPTIONS:

- WOOD FRAME
- 1600 FT^2
- TWO-STORY
- APPROX. SQUARE (25'×32')
- ST. LOUIS DESIGN TEMP = − 5°F
  - AVERAGE TEMP=45°F
  - DEGREE DAYS=4500

Figure E.1

	FEET^x	R	U FACTOR	BTU/HR. 1°ΔF	BTU/DD	BTU/HR. 70°ΔF	BTU/ 70DD
Wall	1620^2	11.9	.084	136	3264	9520	228,480
Ceiling	800^2	21.6	.046	36.8	883.2	At 35°F 1288	30,916
Floor	800^2	21.6	.046	36.8	883.2	1288	30,916
Windows	200^2	(Storm)	.56	112	2688	7840	188,160
Volume	12800^3	$\frac{1.5 \text{ AC}}{\text{HR}}$	(.018)	345.6	8294.4	24192	580,608
				667	16,012	44,128	1,059,080

Figure E.2

The purpose of this appendix is to illustrate, through an example, a straightforward approach to isolating the effect any one energy conserving effort will have on reducing the energy required to provide heating for the winter.

This method is applicable to determining the effect of measures taken to remodel existing dwellings, although the illustration will deal specifically with new construction. New construction was chosen to better show the different options one has for reducing energy requirements—without the constraints of an existing situation.

The example is based on a two-story, wood frame house of 1600 ft² —which is the average size of homes in the United States. (Figure E.1). For computation of the outside wall area it was assumed the house was 25×32 feet with 8 foot high ceilings. For a starting point, an R - 11.9 for the walls and R 21.6 for the ceilings and the floor were assumed as minimum, which is quite good by current standards. Glass area was set at 12.5% of the floor area, (codes set it at 10% of Living Area), and it was assumed that all

glass had storm windows, and all doors were fitted with storm doors. The climate setting for the house is St. Louis, which has a yearly total of 4500 degree days.

These assumptions are carried through a complete analysis to establish a baseline against which further improvements can be gauged. Figure E.2 is a table which describes the heat loss under different conditions for each element involved in heat loss to the outside. It should be noted that the attic above the ceiling and the crawl space beneath the floor are fairly well sealed. This reduces their heat loss to ½ of what it would be at design conditions (35 Δ°F/70 Δ°F). The rate of infiltration is assessed at 1.5 air changes per hour. This is CASE No. 1.

Case No. 1 is broken down to show the % contribution of each element to heat loss. The total BTU/DD is multiplied by the DD/Year to establish the yearly heat loss, or demand. This yearly demand is then divided by 0.7 to establish the amount of energy required at a conversion efficiency of 70% (Figure E.3)—the yearly supply of heat required.

BTU/DD	%	
8294	51.8	INFILTRATION
3264	20.5	EXTERIOR WALL
2688	16.7	WINDOWS
883	5.5	CEILING
883	5.5	FLOOR
16012	100.0	

$$
\begin{array}{r}
16{,}012 \text{ BTU/DD} \\
\times \quad 4{,}500 \text{ DD/WINTER} \\
\hline
72{,}054{,}000 \text{ BTU/WINTER (LOAD)}
\end{array}
$$

At 70% EFF.
103,000,000 BTU SUPPLIED

103×10^6 BTU = 730 Gal No. 2 OIL
= 103,000 CU.FT. of GAS
= 22,240 KW (95% EFF.)

Figure E.3

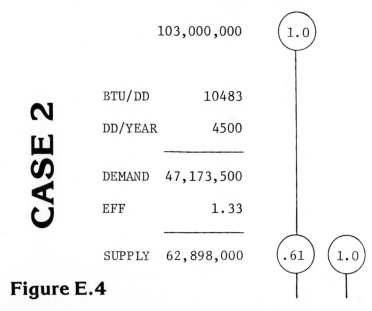

CASE 2

103,000,000	1.0
BTU/DD	10483
DD/YEAR	4500
DEMAND	47,173,500
EFF	1.33
SUPPLY	62,898,000

.61   1.0

Figure E.4

CASE No. 2 represents the first energy conservation step taken. From Case No. 1 it is noticed that over half the energy is lost through infiltration. So this element is chosen as the most important point to begin reducing the load. The goal is to cut infiltration by two-thirds; and it is accomplished by concentration on very good weather stripping and caulking of all openings, by installing casement type windows and air locked (buffered) entries and by supplying fresh air to the furnace through a heat recovery device on the furnace exhaust. These changes all reduce infiltration, and their effect is shown in Figure E.4.

CASE No. 3 represents the next step to be taken. From Case No. 2 it can be seen that the exterior walls now account for almost one-third of the heat loss. However, it is probably better to focus on the windows first, because they represent one-fourth the heat loss and only one-eighth the area of the walls. The heat loss per area is much higher for the glass, so it would probably be more economical to tackle the problem of the glass first. Storm windows, which are slightly better than double insulating glass, is already assumed. Upgrading to triple insulating glass will reduce loss through the glass by one-sixth; this might

BTU/DD	%	
2765	26.3	INFILTRATION
3264	31.1	EXTERIOR WALL
2688	25.6	WINDOWS
883	8.5	CEILING
883	8.5	FLOOR
10483	100.0	

CASE 3

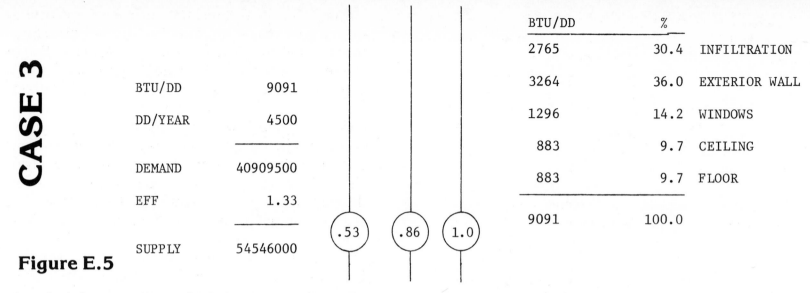

BTU/DD	9091
DD/YEAR	4500
DEMAND	40909500
EFF	1.33
SUPPLY	54546000

(.53) (.86) (1.0)

BTU/DD	%	
2765	30.4	INFILTRATION
3264	36.0	EXTERIOR WALL
1296	14.2	WINDOWS
883	9.7	CEILING
883	9.7	FLOOR
9091	100.0	

Figure E.5

be a lot of extra expense for little gain. However, if insulating shutters are closed over the windows at night, the heat loss through the windows can be cut in half. This is based on using a shutter with 2 inches of styrofoam and on keeping the shutters closed for the average 13.5 hour winter night at an R of 10.56 (Fig. E.5).

For CASE No. 4, the exterior walls appear to be the next element for reduction in heat loss. The suggested improvement would be to place 2 inches of styrofoam on the outside of the studs. This upgrades the R value of the wall to R − 21 + and cuts the heat loss through the walls by more than 40%.

These represent improvements that are accomplishable with additional effort. Trying to reduce losses from Case No. 4 will be quite difficult, since most of the "slack" has been cut out. Once again, infiltration is the major source of heat loss. However, one-half air change an hour is already very good; there is little that can be done that is reasonable to reduce infiltration further in this example (Figure E.6.).

CASE 4

BTU/DD	7678
DD/YEAR	4500
DEMAND	34551000
EFF	1.33
SUPPLY	46068000

(.45) (.73) (.84)

BTU/DD	%	
2765	36.	INFILTRATION
1851	24.	EXTERIOR WALL
1296	17.	WINDOWS
883	11.5	CEILING
883	11.5	FLOOR
7678	100.0	

Figure E.6

CASE No. 5. If further reduction were required, it would probably be important to place some of the house below grade (Figure E.7). The heat loss characteristics of putting three walls and the floor of the first floor below grade are shown in Figure E.8. In Cases 1-4, all heat loss was a function of ambient air temperature. Now, in Case 5, there are elements that experience steady-state heat loss—they are not weather dependent losses.

In order to calculate the heat loss for the season it is necessary to separate the weather dependent and the steady-state portions of heat loss. The weather dependent portion of the BTU/DD is actually the heat loss for each day (BTU/Day), and it is multiplied by 212—the number of days of winter (October 1-May 1). The total for Case 5 is only slightly different than Case 4.

What is advantageous about placing the structure below grade is that the peak load of the structure is markedly reduced. For Case 4, the heat loss during one day at design conditions of −5°F is 475,650 BTU/Day; with Case 5 this becomes 368,044 BTU/Day—almost a 25% reduction. Although the yearly heat losses for Cases 4 and 5 are virtually the same, the reduction of peak load in Case 5 allows a reduction in the amount of heating equipment required and the load is better suited to solar heating, since it stretches the capability of storage a little farther for the sunless periods. Shifting some of the heating load into a steady-state situation means that more of the total seasonal heat requirements occur in early and late winter, where they are easily compensated for by solar energy utilization.

Figure E.7

CASE 5

		BTU/HR. 1°ΔF	BTU/DD	BTU/HR. 70°ΔF	BTU/ 70DD
WALL	ABOVE GRADE	45.8	1101	3211	77064
	BELOW GRADE*	656	15744	656	15744
CEILING		36.8	883	1288	30916
FLOOR	SLAB*	1600	38400	1600	38400
	EDGE	10.9	261	768	18432
WINDOWS		112	1296	7840	90720
VOLUME	INFILTRATION	57.6	1382	4032	96768
		2519.1	59067	19395	368044	TOTALS

*STEADY STATE

59,067 BTU/DD

STEADY STATE	WEATHER DEPENDENT
54,144 BTU/DAY	4,923 BTU/DD
x 212 DAYS/WINTER	x 4500 DD/WINTER
12,153,500 BTU/WINTER	22,153,500 BTU/WINTER

```
           33,632,000 BTU DEMAND/WINTER
     x            1.33 EFF (75%)
           44,730,597 BTU/SUPPLY OF FUEL
```

Figure E.8

APPENDIX F

HEAT TRANSFER

GENERAL

When two objects having different temperatures are put into thermal contact with each other, they eventually reach thermal equilibrium at a common final temperature. The process of attaining thermal equilibrium in such a situation may be discribed as a transfer of heat from one object to the other. That is, it is possible to account for the temperature changes by assuming that the amount of heat lost by one object equals the amount of heat gained by the other object.

The physical basis of heat is energy; a transfer of heat from one object to another corresponds to a transfer of mechanical energy associated with the motions of individual atoms and molecules. Heat is microscopic mechanical energy in transit from one body to another.

SPECIFIC HEAT

What determines how much heat must be added to a body to produce a given temperature increase? Clearly, this depends both on the amount of temperature change and on the quantity of matter contained in the body. Denoting the quantity of heat added to the body by Q, then it can be said

$$Q = mc\Delta T$$

where m is the mass of the object, ΔT the temperature change, and c a proportional factor. In general c is different for different materials, but it depends only on the kind of material, and not on the quantity. It is usually called the specific heat of the material; it is the quantity of heat which must be added to unit mass to increase its temperature by 1°.

LATENT HEAT

When a material undergoes a change of state, as when ice melts or water boils, heat is added without change of temperature. It has been found that a definite quantity of heat per unit mass is associated with such a transformation. For example, at ordinary pressures 80 cal of heat is required to change 1 g of ice at 0°C to liquid water at 0°C. This heat does not change the temperature but only the state of the material. It is called the latent heat of fusion; that is, the latent heat of fusion of water is 80 cal/g. Similarly, to change liquid water at 100°C and atmospheric pressure to steam at the same pressure requires 540 cal for each gram of water; the latent heat of vaporization of water is 540 cal/g. These transformations are particular examples of a class of phenomena called phase transformation.

CONDUCTION

The process of heat transfer which is easiest to describe quantitatively is that of conduction of heat. Whenever two parts of a body are at different temperatures, a spontaneous flow of heat takes place from the region of higher temperature to that of lower temperature. This process is known as conduction of heat. A qualitative understanding of the microscopic basis of conduction of heat may be gained by recalling that heat is microscopic mechanical energy of individual atoms and molecules. When the molecules in one region have, on the average, more kinetic energy than those in the neighboring region, they transfer energy to their neighbors in collisions with them. This transfer of energy takes place, on the average, from a region of higher temperature (and greater molecular motion) to one of lower temperature.

Clearly, in solar energy applications, the idea is to get the energy, Q, which is supplied by the sun and captured by collectors, to flow from the higher temperature of the absorber—across a temperature difference, ΔT,—to the lower temperature of the collection fluid. The absorber and the fluid both have mass and specific heats, and they are in contact with each other for an arbitrary amount of time. The longer they remain in contact with each other the more likely it is that they will equilibrate,—come to rest at the same temperature. When this occurs there is no net transfer of energy from one to the other. This condition would represent the upper limit on time that the two should be in contact with one another, or the minimum flow rate.

CONVECTION

There are two important mechanisms of heat transfer in addition to conduction; one of these is convection. In convection, heat is transferred by mass motion of the material. In a hot-water heating system, for example, heat is transferred from the furnace to the room being heated by flow of hot water from the furnace into the radiator, then by conduction from the water to the radiator and to the room. It is much more difficult to make quantitative statements about heat transfer by convection than about conduction, and it will not be attempted here. In situations in which matter is free to move, as when fluids of any kind are involved, convection may be an important effect. Within a solid, convection does not occur.

RADIATION

A third mechanism of heat transfer is radiation. In radiation, thermal energy is transferred through emission and absorption of electromagnetic radiation by matter. For this reason radiation involves a somewhat generalized concept of heat transfer. The rate of heat loss by radiation from a hot body whose absolute temperature is T is proportional to $T^4-T_0^4$, where T_0 is the absolute temperature of the surroundings. The result of this very strong dependence on temperature is that at room temperature radiation is usually negligible as a mode of heat transfer, but at high temperatures, such as in the filament of an incandescent light bulb, it is usually far more important than either of the other mechanisms discussed.

The effect that any surface has on incident radiation is expressed:

$$a + t + r = 1$$

a - absorptivity
t - transmissivity
r - reflectivity

for each wavelength of radiation. Transmissivity is the ability of the surface to let radiation pass through it—like light through a window. Reflectivity is the ability of a surface to reflect radiation back to its surroundings.

If a surface does not pass radiation through it, it is said to be opaque and the transmissivity of the surface is zero:

$$a + r = 1.$$

In addition, for each wavelength of radiation, the absorptivity (a) is equal to the emissivity (e):

$$a = e.$$

The typical, flat black surface that is used in flat plate

solar collectors has an absorptivitiy of:

a = .95 Flat Black.

this means that very little energy is reflected back to the surroundings. It also means that the emissivity of the surface is:

e = .95

which is also very high.

The emissive power (E) from a hot body is equal to:

$$E = ekT^4.$$

T - absolute temperature
°K = °C + 273
°R = °F + 460

k - Stefan - Boltzmann constant

However the hot body is receiving (absorbing) radiation from its surroundings in a similar fashion based on the temperature of the surroundings. The actual heat loss (Q_L) is a function of the difference in temperature between the hot body and its surroundings:

$$Q_L = ek(T_a^4 - T_s^4).$$

T_a - Temp. absorber
T_s - Temp. surroundings

Because of this fourth power relationship and the fact that it represents very high heat loss at high temperatures, it is important to try to operate a collector system at as low a temperature as is reasonably possible and still produce the temperatures required. Going to selective surfaces is another way to deal with the high radiative loss. Selective surfaces are discussed in Chapter 3.

It should be emphasized that a difference in temperature between two bodies is necessary for heat transfer. When a body is in thermal equilibrium with its surroundings, the body and the surroundings have the same temperature; in this case there is no net transfer of heat between the two. Conversely, in steady-state heat flow, the system is not in thermal equilibrium; different points have different temperatures, but heat transfer occurs in such a way that the temperature at each point is constant in time.

APPENDIX G

CONVERSION FACTORS

The actual heat content of fuels varies according to the grade, type of refinery product and chemical composition. This table gives some average heat values for common fuels. Actual heat values vary from deposit to deposit because of differences in chemical composition.

Average values of energy (heat) content of fuels:

1 metric ton coal = 7.06×10^6 kcal, 2.8×10^7 BTU
1 ton coal   = 6.60×10^6 kcal, 2.6×10^7 BTU
1 barrel oil   = 1.46×10^6 kcal, 5.79×10^6 BTU
1 gallon gasoline = 3.48×10^4 kcal, 1.38×10^5 BTU
1 cubic foot gas   = 261 kcal,   1035 BTU

CONVERSION CONSTANTS

A	EQUALS	B
British Thermal Unit (BTU)		1055.1 joules
		252 cal
		0.293 w·hr
British Thermal Unit/pound (BTU/ft² · hr)		1.876 cal/gm
British Thermal Unit/foot² · hr (BTU/ft² · hr)		3.15 w/m²
*calories (cal)		4.184 joules
		4.184 watt · sec
		0.00397 BTU
*calories/gram (cal/g)		1.798 BTU/lb
Centigrade (C)		$(1.8 \, (°C) + 32)°F$
centimeter (cm)		0.394 in
cubic foot (cu ft)		7.48 gal
		0.028 m³
cubic meters (cu m)		35.314 ft³
Fahrenheit (F)		$(°F - 32)(0.555)°C$
foot		.3048 meters
gallon (gal)		0.1336 ft³
		3.784 liter
		8.345 lbs of water

horsepower (hp)		0.745 kw
		2544 BTU/hr
inch (in)		2.54 cm
joule		0.2389 cal
		.00095 BTU
kilocalories (kcal)		3.97 BTU
		1000 cal
kilogram (kg)		2.2 lbs.
kilogram/meter² (kg/m²)		0.205 lb/ft²
kilowatt (kw)		1.34 hp
kilowatt · hour (kwhr)		3413 BTU
langleys		3.69 BTU/ft²
		1.00 cal/cm²
langleys/minute		0.0698 w/cm²
liter (1)		0.264 gal
		0.001 m³
		1.06 qt
meter (m)		3.28 ft
pound (lb)		454 g
quart (qt)		0.946 liter
square foot (ft²)		0.0929 m²
square meter (m²)		10.8 ft²
therm		100,000 BTU
		25200 kcal
ton		907 kg
		2000 lb
watt		3.413 BTU/hr
		1.0 joule/sec
		.239 cal/sec
watts/square centimeter (w/cm²)		14.3 langleys/min
		3171 BTU/(hr · ft²)
watts/meter² (w/m²)		0.317 BTU/(hr · ft²)

*Note that the term "calorie that is used in discussions of food is equal to a kilocalorie in terms of its energy value.

APPENDIX H

BIBLIOGRAPHY

AIA Research Corporation for the Department of Housing and Urban Development, **Solar Dwelling Design Concepts**, (U.S. Government Printing Office, 1976).

Altman, P.I., "Conservation and Better Utilization of Electric Power by Means of Thermal Energy Storage and Solar Heating," Interim NSF Report, NTIS Doc. No. PB210359 (Oct. 1971).

Anderson, B., "Solar Energy and Shelter Design," Master of Architecture Thesis MIT, (January, 1973) 151 p.

Anderson, Bruce with Michael Riordan, **The Solar Home Book**, (Harrisville, New Hampshire: Cheshire Books, 1976).

ASHRAE, **Guide and Data Book: Fundamentals and Equipment** (New York: ASHRAE, 1966).

ASHRAE, **Handbook and Product Applications: 1974 Applications** (New York: ASHRAE, 1974), pp. 59.4-59.5 Adapted from tables.

ASHRAE, **Handbook and Product Directory, 1978 Applications**, (New York, American Society of Heating, Refrigerating and Air-conditioning Engineers, Inc. 1978).

ASHRAE, **Low Temperature Engineering Applications of Solar Energy** (New York: ASHRAE, 1967).

Backus, Charles E., "Photovoltaic Program in the United States," Paper presented by Symposium **Solar Utilization Now** (Arizona State University, January, 1975).

Baer, Steve, "Solar House," **Alternative Sources of Energy** (Kingston, N.Y.: Alternative Sources of Energy, 1974), pp. 39-41.

Barnes, J., **et. al., Wall Design Handbook** (St. Louis: School of Architecture, Washington University, 1974).

Berg, Charles A., "Energy Conservation through Effective Utilization," **Science**, Vol. 181 (July 13, 1973), pp. 128-138.

Bliss, R.W., Jr., "The Derivations of Several 'Plate Efficiency Factors' Useful in the Design of Flat-Plate Solar Heat Collectors," **Solar Energy** (Vol. 3, No. 4, 1958), pp. 55-64.

Boaz, J.N., editor, **Architectural Graphic Standards**, 6th edition (New York: John Wiley and Sons, 1970), p. 313.

Boer, K. W., "Part 1: The Solar One Conversion System," **Solar One: First Results** (Newark, Delaware: Institute of Energy Conversion, University of Delaware, 1974).

Brinkworth, B.J., **Solar Energy for Man** (New York: John Wiley and Sons, 1970), p. 313.

Burchberg, H., **et al.**, "Performance Characteristics of Rectangular Honeycomb Solar-Thermal Converter," **Solar Energy**, Vol. 13, (1971), pp. 193-221.

Byers, H.R., **General Meteorology** (New York: McGraw-Hill Book Co., 1959).

Chahroudi, D., and S. Wellesley-Miller, "Sun, Wind and Shelter: Buildings as Organisms," Department of Architecture, MIT (February, 1974), 16 pages.

Clarkson, Clarence W. and James S. Herbert, "Transparent Glazing Media for Solar Energy Collectors," International Solar Energy Society, U.S. Section Annual Meeting (Aug. 1974), 13 pp.

Close, D.J., "Solar Air Heaters," **Solar Energy** (Vol. 7, No. 3, 1963), p. 117-124.

Davis, Paul, "To Air Is Human: Some Humanistic Principles in the Design of Thermosiphon Air Heaters," **Proceedings of the Passive Solar Heating and Cooling Conference**, (May 18-19, 1976).

Daniels, Farrington, **Direct Use of the Sun's Energy** (New York: Ballantine Books, 1975).

Decker, Jack, "Demonstration of Solar Energy Utilization," Brochure of **Solar Systems, Inc.**, Tyler, Texas.

U.S. Department of Commerce, Social and Economic Statistics Administration, Bureau of the Census. Photocopy of unpublished computer printout Section 203, Existing Construction 1973.

Department of the Environment, **Thermal Insulation of Buildings**, (London: Her Majesty's Stationery Office, 1971).

Department of Housing and Urban Development, Office of International Affairs, "Information Series 24" (August 27, 1973), p.4.

Dubin, F.S., "Energy Conservation through Building Design and a Wiser Use of Electricity," Paper at the Annual Conference of the American Public Power Association (San Francisco, California, June, 1972).

Dubin, F.S. "Notes and Comments on Solar Energy," **Actual Specifying Engineer**, (Nov., 1973), p. 69.

Duffie, J.A. and W.A. Beckman, **Solar Energy Thermal Processes** (New York: John Wiley and Sons, 1975.)

Eaton, C.B. and H.A. Blum, "The Use of Moderate Vacuum Environments as a Means of Increasing the Collection Efficiencies and Operating Temperatures of Flat-Plate Solar Collectors," **Technical Program and Abstracts** (Fort Collins, Colorado: ISES Annual Meeting, Aug., 1974), p. 12.

Edmund Scientific Co., Catalog 751.

Engebretson, C.D. and N.G. Ashar, "Progress in Space Heating with Solar Energy," ASME paper No. 60-WA-88 (December, 1960).

Fitch, J.M., **American Building: The Environmental Forces that Shape It** (Boston: Houghton Mifflin Company, 1972), 349 pp.

General Services Administration, **Energy Conservation Guidelines for Office Buildings** Dubin: Mindell Bloom Associates, (Jan., 1974), pp. 9-14.

Glaser, P.E., "Fueling the Earth with the Sun." **Environmental Design**, (Summer/Fall, 1973), pp. 18-20.

Hammond, A.L., "Individual Self-Sufficiency in Energy," **Science**, Vol. 184, (April 19, 1974), p. 281.

Hay, Harold, "New Roofs for Hot Dry Regions," **Ekistics**, Vol. 183 (Feb. 1971), pp. 158-64.

Hay, H.R. and J.I. Yellot, "A Naturally Air-Conditioned Building," **Mechanical Engineering** (Jan. 1970), pp. 19-25.

Hildebrandt, A.F., **et al**., "Large-Scale Concentration and Conversion of Solar Energy," **Transactions of the American Geophysical Union** Vol. 53, No. 7, (July, 1972), pp. 684-92.

Hottel, H.C. and B.B. Woertz, "The Performance of Flat-Plate Solar Heat Collectors," **Transactions of the ASME** Vol. 64, pp. 91-104 (Feb. 1942).

Hutchinson, F.W., "The Solar House: Analysis and Research," **Progressive Architecture** (May 1947), pp. 90-94.

International Conference of Building Officials, **Uniform Building Code**, 1973 ed., p. 88.

Keys, John, **Harnessing the Sun** (Dobbs Ferry, New York: Morgan and Morgan, 1975).

Kobayashi, Takao and Stephen L. Sargent, "A Survey of Breakage Resistant Materials for Flatplate Solar Collector Covers," International Solar Energy Society, U.S. Section Annual Meeting (Aug. 1974), p. 12.

Kreider, J. F. and F. Kreith, **Solar Heating and Cooling**, (Washington, D.C., Hemisphere Publishing Corporation, 1975).

Kreider, J. F. and G. Steward, "Stationary Reflector/Tracking Absorber Solar Concentrator," **Technical Program and Abstracts** (Fort Collins, Colorado: ISES Annual Meeting, Aug., 1974), p. 24.

Kuzay, T.M., **et al.**, **Solar One: First Results** (Newark, Delaware: Institute of Energy Conversion, University of Delaware, 1974).

Labs, Ken, The Architectural Use of Underground Space: Issues and Applications (St. Louis: Washington University, Masters of Architecture Thesis, May, 1975), p. 36.

Lalude, O., and H. Buchberg, "Design and Application of Honeycomb Porous-Bed Solar-Air Heaters," **Solar Energy**, Vol. 13 (1971), p. 223-242.

Langer, D., "Characteristics of Cadmium Sulfide Photovoltaic Cells," **Solar and Aeolian Energy**, Proceedings of the International Seminar on Solar and Aeolian Energy, Sounion, Greece, Sept. 4-15, 1961 (New York: Plenum Press, 1964), pp. 234-235.

Leckie, Jim, **et al.**, **Other Homes and Garbage**, (San Francisco, Sierra Club Books, 1975).

Liu, B.Y.H. and R.C. Jordan, "Chapter I: Availability of Solar Energy for Flat-Plate Solar Heat Collectors," **Low Temperature Engineering Applications of Solar Energy** (New York: ASHRAE, 1967), 18 p.

Liu, B.Y.H. and R.C. Jordan, "The Long Term Average Performance of Flat-Plate Solar Energy Collectors," **Solar Energy** (Vol. 7, No. 2, 1963), p. 53-70.

Local Climatological Data, U.S. Department of Commerce, National Oceanic and Atmospheric Administration, St. Louis, Mo.

Lof, G.O.G., **et al.**, "Design and Construction of a Residential Solar Heating and Cooling System," NTIS Doc. No. PB237042 (August, 1974).

Lof, G.O.G. and R.A. Tybout, "Cost of House Heating with Solar Energy," **Solar Energy** (Vol. 14, 1973), p. 253-78.

Londe-Parker-Michels, Inc., "Early Field Experience and Cost Effectiveness of a Passive Solar Heating Retrofit," **Proceedings of the 3rd National Conference on Technology for Energy Conservation**, (Tucson, Arizona, January 23-25, 1979).

Londe-Parker-Michels, Inc., Final Task Report on Residential Redesign, Phase II of Building Energy Performance Standards, subcontract report to the AIA/Research Corporation, Sept. 25, 1978.

Lowry, W. P., "The Climate of the Cities," **Scientific American**, (August, 1967).

Mattingly, G.E. and E.F. Peters, "Wind and Trees—Air Infiltration Effects on Energy in Housing," (Center for Environmental Studies, Report No. 20, May, 1975, Princeton University, NSF Grant No. 11237).

Mazria, E., S. Baker and F. Wessling, "Predicting the Performance of Passive Solar Heated Buildings," **Proceedings of the 2nd National Passive Solar Conference: Volume II**, (March, 1978, Philadelphia).

McCullagh, James C., Editor, **The Solar Greenhouse Book**, (Emmans, Pa., Rodale Press, Inc., 1978).

McGuiness, W.J. and B. Stein, **Mechanical and Electrical Equipment for Buildings**, 5th edition (New York: John Wiley and Sons, Inc. 1971), 1011 p.

Miller, Charles A., "A Low-Cost Solar Cell Is Here!", **Mechanix Illustrated** (April, 1978).

Ministry of Public Building and Works, **Metrication in the Construction Industry, No. 2**. (London: Her Majesty's Stationary Office, 1970).

Moorcraft, C., "Solar Energy in Housing," **Architectural Design** Vol. 43 (Oct. 1973), p. 634.

Moore, G.L., "Sizing of Solar Energy Storage Systems Using Local Weather Records," 1975.

Moore, S.W., **et al.**, "Design and Testing of a Structurally Integrated Steel Solar Collector Unit Based on Expanded Flat Metal Plates," International Solar Energy Society, U.S. Section Annual Meeting (Aug. 1974), 34 p.

The Mother Earth News, **Handbook of Homemade Power** (New York: Bantam Books, May, 1974), 374 p.

NASA, "Efficiency Increased in New Solar Cell: A Concept," **Tech Brief B74-10090**, Langley Research Center (August, 1974).

NASA, "A Practical Solar Energy Heating and Cooling System," **Technical Brief B73-10156**, Marshall Space Flight Center (May, 1973), p. 12.

NASA, **TERRASTAR** Final Report GR-129012 (Sept. 1973).

NSF/NASA Solar Energy Panel, **Solar Energy as a National Resource**, U.S. G.P.O. Doc. 3800 - 00164, p. 56.

National Bureau of Standards, "Interim Performance Criteria for Solar Heating and Combined Heating/Cooling Systems and Dwellings" (January, 1, 1975).

O'Connor, E., "Solar Energy—How Soon?" **Chemtech** (May, 1974), pp. 264-67.

Olgay, A. and M. Telker, "Solar Heating for Houses," **Progressive Architecture** (March, 1959), pp. 195-203.

Olgay, V., **Design with Climate** (Princeton: Princeton University Press, 1963).

Peavy, B. A., F. J. Powell, and D. M. Burch, **Dynamic Thermal Performance of an Experimental Masonry Building**, Building Science Series 45, U.S. Department of Commerce, National Bureau of Standards, (July, 1973).

Rabl, A. and C.E. Nielsen, "Solar Ponds for Space Heating," International Solar Energy Society, U.S. Section Annual Meeting (Aug., 1974), 41 pp.

Ramsey, C. G. and H. R. Sleeper, **Architectural Graphic Standards, 6th Edition**, (New York: John Wiley and Sons, Inc., 1970).

Satcunanathan, S. and S. Deonarine, "A Two-Pass Solar Air Heater," **Solar Energy** Vol. 15, (1973), pp. 41-49.

Schiff, M., "Direct Gain Passive Solar Design in an Extreme Climate," **Proceedings of the 2nd National Passive Solar Conference: Volume I**, (March 1978, Philadelphia).

Shurcliff, W.A., "Solar Heated Buildings: A Brief Survey," 5th edition, (August 17, 1974).

Steadman, P.J., **Energy, Environment, and Building** (New York: Cambridge University Press, 1975), 287 p.

Spanides, A.G., **Solar and Aeolian Energy** (New York: Plenum Press, 1961), p. 118. Published proceedings of International Seminar on Solar and Aeolian Energy held at Sounion, Greece, (Sept., 1961).

Steinhart, C., and J. Steinhart, **Energy: Sources, Use and Role in Human Affairs** (North Scituate, Mass.: Duxbury Press, 1974), 362 p.

Sweet's Architectural Catalogs (New York: McGraw Hill, 1971).

Tabor, H., "Chapter IV: Selective Surfaces for Solar Collectors," **ASHRAE, Low Temperature Engineering Applications of Solar Energy** (New York: ASHRAE, 1967), p. 41-52.

Telkes, M., "Solar House Heating—A Problem of Heat Storage," **Heating and Ventilating** (May, 1947), pp. 68-75.

Threlkeld, J.L., **Thermal Environmental Engineering** Englewood Cliffs, New Jersey: Prentice-Hall Inc., 1962).

Tiemann, H.D., **Wood Technology** (New York: Pitman Publishing Co., 1942).

Trewartha, G.T., **An Introduction to Climate** (New York: McGraw-Hill Book Company, 1968).

Ward, D. S. and G. O. G. Lof, "Design and Construction of a Residential Solar Heating and Cooling System," **Solar Energy**, (Vol. 17, No. 1, 1975).

Weast, R.C., Editor, **CRC Handbook of Chemistry and Physics 49th Edition** (Cleveland, Ohio: The Chemical Rubber Co., 1968-1969).

Weast, R.C., Editor, **CRC Handbook of Chemistry and Physics 54th Edition** (Cleveland, Ohio: CRC Press, 1973).

Whillier, A., "Thermal Resistance of the Tube-Plate Bond in Solar Heat Collectors," **Solar Energy**, Vol. 8, No. 3 (1964) pp. 95-98.

Woodliff, Frank III, "Towards Solarchitecture in Utah, Part 2," **Utah Architect** (Autumn, 1974), p. 18-21.

Wright, J.R. and P.R. Achenbach, **Scientific American Roundtable on Energy Conservation in Buildings** (August 29, 1973).

Yanda, Bill and Rick Fisher, **The Food and Heat Producing Solar Greenhouse Design, Construction and Operation**, (Santa Fe, John Muir Publications, 1976).

Yellot, J.I., "Utilization of Sun and Sky Radiation for Heating an Cooling of Buildings," **ASHRAE Journal** (Dec. 1973), pp. 31-42.

Young, H.D., **Fundamentals of Mechanics and Heat** (New York: McGraw-Hill, 1964), 638 pp.

APPENDIX I

SOLAR COLLECTION
SIZING PROCEDURE

SOLAR COLLECTOR SIZING PROCEDURE
Normalization Technique

INCIDENT CLEAR DAY INSOLATION — SOUTHFACING SURFACE

Surface Daily Totals (BTU/sq. ft.)
40° North Latitude
From ASHRAE 1974 Applications Handbook, page 59.4 of Chapter 59.

	Sun Tracking	0°	40°	50°	60°	90°
Jan. 21	2182	948	1810	1906	1944	1726
Feb. 21	2640	1414	2162	2202	2176	1730
Mar. 21	2916	1852	2330	2284	2174	1484
Apr. 21	3092	2274	2320	2168	1956	1022
May 21	3160	2552	2264	2040	1760	724
June 21	3180	2648	2224	1974	1670	610
July 21	3062	2548	2230	2006	1728	702
Aug. 21	2916	2244	2258	2104	1894	978
Sep. 21	2708	1788	2228	2182	2074	1416
Oct. 21	2454	1348	2060	2098	2074	1654
Nov. 21	2128	942	1778	1870	1908	1686
Dec. 21	1978	782	1634	1740	1796	1646

TABLE A

SLOPE / HORIZONTAL RATIO

40° North Latitude

	40°	50°	60°	90°
Jan. 21	1.91	2.01	2.05	1.82
Feb. 21	1.53	1.56	1.54	1.22
Mar. 21	1.26	1.23	1.17	.80
Apr. 21	1.02	.95	.86	.45
May 21	.89	.80	.69	.28
June 21	.84	.75	.63	.23
July 21	.88	.79	.69	.28
Aug. 21	1.01	.94	.84	.44
Sep. 21	1.25	1.22	1.16	.79
Oct. 21	1.53	1.56	1.54	1.23
Nov. 21	1.89	1.98	2.03	1.79
Dec. 21	2.09	2.22	2.30	2.10

Table A is produced by dividing the incident clear day insolation on the different tilted surfaces by the insolation on a horizontal surface. This normalizes the data to a horizontal surface. This is done because the long-term weather data that is kept by the Department of Commerce is measured on and recorded for horizontal surfaces.

TABLE B

AVERAGE DAILY INSOLATION ON A HORIZONTAL SURFACE

BTU/Square Foot   St. Louis, Missouri

OCT	NOV	DEC	JAN	FEB	MAR	APR
980	634	529	618	834	1084	1392

Table B is the measured long-term average insolation available on a horizontal surface in the St. Louis area. (Columbia, Mo. data that has been modified for the St. Louis microclimate.)

TABLE C

INSOLATION LEVELS NORMALIZED TO JANUARY INSOLATION

40°	1.53	1.89	2.09	1.91	1.53	1.26	1.02	A
COLL. TILT	1499	1189	1105	1180	1276	1366	1420	AXB

THESE BTU'S ARE NORMALIZED TO JANUARY

1.27	1.02	.94	1.0	1.08	1.16	1.2	C/40°

60°	1.54	2.03	2.30	2.05	1.54	1.17	.86	A
COLL. TILT	1509	1287	1217	1267	1284	1268	1197	AXB

1.19	1.02	.96	1.0	1.01	1.00	.94	C/60°

90°	1.23	1.79	2.1	1.82	1.22	.80	.45	A
COLL. TILT	1205	1135	1111	1125	1017	867	626	AXB

1.07	1.01	.99	1.0	.9	.77	.56	C/90°

These tables adjust the energy available on a horizontal surface to that available on a tilted surface. This is accomplished by multiplying **Table B**, "Avg. Daily Insolation on a Horizontal Surface" by **Table A**, "Tilt Multiplying Tables," then the avg. daily energy for each month is divided by the avg. daily energy for January. This normalizes the available solar energy to the worst heating case in January.

TABLE D
(St. Louis, Mo.)

OCT	NOV	DEC	JAN	FEB	MAR	APR
224	611	947	1064	854	655	306

TOTAL: 4661

Table D is the long-term average Degree Days for each winter month.

TABLE E
Seasonal Performance Analysis

Collectors are to satisfy 50% of January heating load of 1064 DD. Therefore, a collector area is required that will supply the heat need to cover 532 DD in January.

C/40°
This tilt and size will cover 3296 DD of the full 4661 DD.

$$\frac{3296}{4661} = 70.7\%$$

OCT	NOV	DEC	JAN	FEB	MAR	APR
			532 X			
$\frac{1.27}{(224)}$	$\frac{1.02}{543}$	$\frac{.94}{500}$	$\frac{1.0}{532}$	$\frac{1.08}{574}$	$\frac{1.16}{617}$	$\frac{1.2}{(306)}$

C/60°
This tilt and size will cover 3185 DD of the full 4661 DD.

$$\frac{3185}{4661} = 68.3\%$$

OCT	NOV	DEC	JAN	FEB	MAR	APR
			532 X			
$\frac{1.19}{(224)}$	$\frac{1.02}{543}$	$\frac{.96}{511}$	$\frac{1.0}{532}$	$\frac{1.01}{537}$	$\frac{1.0}{532}$	$\frac{.94}{(306)}$

C/90°
This tilt and size will cover 3007 DD of the full 4661 DD.

$$\frac{3007}{4661} = 64.5\%$$

OCT	NOV	DEC	JAN	FEB	MAR	APR
			532 X			
$\frac{1.07}{(224)}$	$\frac{1.01}{537}$	$\frac{.99}{527}$	$\frac{1.0}{532}$	$\frac{.9}{479}$	$\frac{.77}{410}$	$\frac{.56}{298}$

Table E illustrates the calculations to determine the overall winter performance of different tilts if the system could supply 50% of the heat required for January.

TABLE F
Collector Area Determination

BASED ON SATISFYING 50% OF THE JANUARY HEATING LOAD

	40°	60°	90°	
COLLECTOR TILT				
HORIZONTAL JANUARY INSOLATION (TABLE B) BTU/SQ.FT. – DAY (AVG)	618	618	618	
		X		
SLOPE/HORIZONTAL RATIO (TABLE A)	1.91	2.05	1.82	AVG. JANUARY
	1180	1267	1125	BTU/SQ.FT.–DAY ON A TILTED SURFACE

Table F calculates the solar energy that falls on each square foot of surface on the average January day for different collector tilts.

To accurately compare the different tilts, the methods used in **Table E** are used to adjust the performance of each tilted surface so that they all provide 70% of the winter heating load. An iterative process was used to find the new Degree Days to be satisfied in January at each tilt:

	40°	60°	90°	
DD/JAN FOR 70% PERFORMANCE	532	553	590	
	←———— ÷31 ————→			DAYS IN JANUARY
AVG DD/DAY	17.2	17.8	19.0	

An average day of sunshine should be able to satisfy the average daily heating Degree Day associated with each tilt.

Question: How many square feet of collector required?

	40°	60°	90°	
FOR 1000 BTU/DD OF BLDG. HEAT LOSS SOLAR SUPPLIES	17200	17800	19000	BTU/AVG. JAN. DAY
AVAILABLE INSOLATION FROM TABLE F	1180	1267	1125	BTU/SQ.FT.AVG. JAN. DAY
	14.57	14.05	16.89	SQ.FT.
	←——ACTIVE——→		PASSIVE	
DIVIDE BY EFFICIENCY	.33	.33	.5	
	43.7	42.2	33.8	SQ.FT./(1000 BTU/DD)

TABLE G

Table G for every 1000 BTU/DD, then, each tilt will require a proportional increase in collector area to provide a 70% solar heated house.

		40°	60°	90°	
		←——— ACTIVE ———→		PASSIVE	
1000	BTU/DD	43.7	42.2	33.8	SQUARE FEET OF COLLECTOR REQUIRED TO SATISFY VARIOUS LOADS.
5000	"	218	211	169	
10000	"	437	422	338	
15000	"	656	633	507	
20000	"	874	844	676	
25000	"	1093	1055	845	
30000	"	1311	1266	1014	

NOTE: Active or passive systems can be 40°, 60°, 90°, or any tilt angle and the appropriate **Table G** can be generated by simply using the correct system efficiency in the preceeding step.

INDEX

(see also Thermal Storage)
Sun, The, 23
Systems:
    capabilities, 132
    efficiencies, 131
    types:
        air collection · air distribution, 99
        air collection · water distribution, 105
        water collection · air distribution, 101
        water collection · water distribution, 105

T

Temperature Swing:
    passive mass control, 44, 46
Thermal Expansion:
    flat plate collector materials, 63
    wood, 13
Thermal Storage, 89
    active, 92
        heat of fusion, 95
        specific heat, 92
            rocks, 94
            water, 93
    dynamics, 113
    passive, 90

V

Value Engineering, 5
Ventilation, 15

W

Walls, 8
Weather Data, 33
    (see also Climate Data)
Windows:
    energy conservation, 5, 8, 12
    passive collectors, 43